DEFEATING
BURGLAR ALARMS

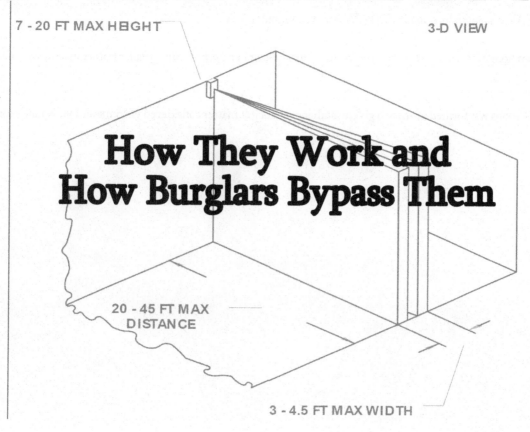

7 - 20 FT MAX HEIGHT

3-D VIEW

How They Work and
How Burglars Bypass Them

20 - 45 FT MAX
DISTANCE

3 - 4.5 FT MAX WIDTH

Defeating Burglar Alarms: How They Work and How Burglars Bypass Them

This edition first published in the United States in 2010 by Warcry Communications

ISBN 978-0-9842844-3-6

Defeating Burglar Alarms, introduction © 2010 by Warcry Communications

Warcry Communications was formed in 2010 to give an audience to subject matter considered extreme and viewpoints considered unpopular.

DEFEATING

BURGLAR ALARMS

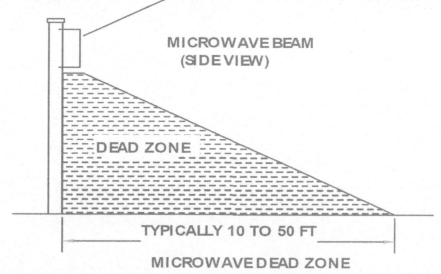

MICROWAVE BEAM
(SIDE VIEW)

DEAD ZONE

TYPICALLY 10 TO 50 FT

MICROWAVE DEAD ZONE

Typical Defeat Measures: The detaching/cutting of an opening in the window or the removal of a window pane (with or without a sensor mounted on it) can bypass the sensor. The break frequency can be distorted by muffling the sound of the breaking glass reducing the potential for the "correct" frequency registered by the sensor.

INTRODUCTION

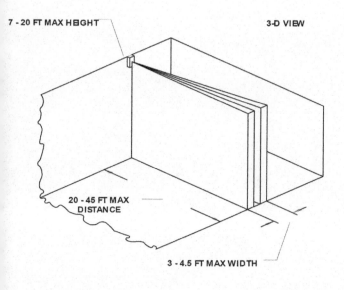

7 - 20 FT MAX HEIGHT   3-D VIEW

20 - 45 FT MAX DISTANCE

3 - 4.5 FT MAX WIDTH

Defeating Burglar Alarms was first published as a military report under the title Perimeter Security Sensor Technologies Handbook. The report analyzed the most commonly used burglar alarms and their weaknesses. This book provides the best available explanation of the technological principles and common defeat measures for all alarms in popular use.

We hope this book gives the layperson a better understanding of the previously arcane technology of burglar alarms. Only with this understanding will those wishing to protect themselves intelligently choose the alarms system best for them.

This book is intended for informational purposes only. Neither the authors nor the publisher assumes any responsibility for the use or misuse of the information contained in this book.

- The editors

TABLE OF CONTENTS

SECTION ONE

INTRODUCTION

1.1 GOAL

This Handbook is intended to be used as a sensor selection reference during the design and planning of perimeter security systems. The Handbook contains a compendium of sensor technologies that can be used to enhance perimeter security and intrusion detection in both permanent and temporary installations and facilities.

1.2 ORGANIZATION

The Handbook is organized into three sections. Section One includes this Overview of a dozen factors to be considered prior to selecting a suite of perimeter detection sensors. Section Two consists of a description of each of the twenty-eight (28) Detection Sensor technologies discussed in the Handbook, including Operating Principles, Sensor Types/Configurations, Applications and Considerations, and Typical Defeat Measures.

1.3 OPERATIONAL REQUIREMENTS

The application of security measures is tailored to the needs and requirements of the facility to be secured. The security approach will be influenced by the type of facility or material to be protected, the nature of the environment, and the client's previous security experience and any perceived threat. These perceptions form the basis for the user's initial judgment, however, these perceptions are rarely sufficient to develop an effective security posture. The nature and tempo of activity in and around the site or facility, the physical configuration of the facility/complex to be secured, the surrounding natural and human environment, along with the fluctuations and variations in the weather, as well as new or proven technologies are all factors which should be considered when planning a security system.

In addition to the large variety of permanent Federal and State facilities located within the confines of the United States that require perimeter security, there is a family of American military, humanitarian, diplomatic and peacekeeping complexes overseas, many of which, although transitory in nature require a dynamic and creative approach to the challenge of perimeter security. Many of the technologies discussed in this handbook can, with some adaptation, be applied to these situations. Typical examples of these complexes include: logistic depots, ship and aircraft unloading and servicing facilities, vehicle staging areas, personnel billeting compounds, communications sites and headquarters compounds. Although the personnel and vehicle screening challenges at each site will vary with the nature of the environment and the potential threat, the role of perimeter security will be similar in all cases.

Basically stated, the role of a perimeter security system is fourfold: deter, detect, document and deny/delay any intrusion of the protected area or facility. In the case of American facilities and complexes located in foreign countries, this challenge is further complicated when U.S. forces cannot patrol or influence the environment beyond the immediate "fenceline". In situations such as these, the area within the fenceline (the Area of Responsibility - AOR), should be complemented by an area of security surveillance beyond the fence, (preferably a cordon sanitaire) wherein the perimeter, from an early warning perspective is extended outward. This is particularly essential in situations where the host government security forces cannot provide a reliable outer security screen, or the area to be secured abuts a built-up industrial, business, public or residential area.

1.4   SYSTEM INTEGRATION

The integration of sensors and systems is a major design consideration and is best accomplished as part of an overall system/installation/facility security screen. Although sensors are designed primarily for either interior or exterior applications, many sensors can be used in both environments. Exterior detection sensors are used to detect unauthorized entry into clear areas or isolation zones that constitute the perimeter of a protected area, a building or a fixed site facility. Interior detection sensors are used to detect penetration into a structure, movement within a structure or to provide knowledge of intruder contact with a critical or sensitive item.

1.5   DETECTION FACTORS

Six factors typically affect the Probability of Detection (Pd) of most area surveillance (volumetric) sensors, although to varying degrees. These are the: 1) amount and pattern of emitted energy; 2) size of the object; 3) distance to the object; 4) speed of the object; 5) direction of movement and 6) reflection/absorption characteristics of the energy waves by the intruder and the environment (e.g. open area, shrubbery, or wooded).

Theoretically, the more definitive the energy pattern, the better. Likewise, the larger the intruder/moving object the higher the probability of detection. Similarly, the shorter the distance from the sensor to the intruder/object, and the faster the movement of the intruder/object, the higher the probability of detection. A lateral movement that is fast typically has a higher probability of detection than a slow straight-on movement. Lastly, the greater the contrast between the intruder/moving object and the overall reflection/absorption characteristics of the environment (area under surveillance), the greater the probability of detection.

1.6   SENSOR CATEGORIES

Exterior intrusion detection sensors detect intruders crossing a particular boundary or entering a protected zone. The sensors can be placed in clear zones, e.g. open fields, around buildings or along fence lines. Exterior sensors must be resilient enough not only to withstand outdoor weather conditions, such as extreme heat, cold, dust, rain, sleet and snow, but also reliable enough to detect intrusion during such harsh environmental conditions.

Exterior intrusion sensors have a lower probability of detecting intruders and a higher false alarm rate than their interior counterparts. This is due largely to many uncontrollable factors such as: wind, rain, ice, standing water, blowing debris, random animals and human activity, as well as other sources to include electronic interference. These factors often require the use of two or more sensors to ensure an effective intrusion detection screen.

Interior intrusion detection sensors are used to detect intrusion into a building or facility or a specified area inside a building or facility. Many of these sensors are designed for indoor use only, and should not be exposed to weather elements.

Interior sensors perform one of three functions: (1) detection of an intruder approaching or penetrating a secured boundary, such as a door, wall, roof, floor, vent or window, (2) detection of an intruder moving within a secured area, such as a room or hallway and , (3) detection of an intruder moving, lifting, or touching a particular object.

Interior sensors are also susceptible to false and nuisance alarms, however not to the extent of their exterior counterparts. This is due to the more controlled nature of the environment in which the sensors are employed.

1.7   TECHNOLOGY SOLUTIONS

With the advent of modern day electronics, the flexibility to integrate a variety of equipment and capabilities greatly enhances the potential to design an Intrusion Detection System to meet specific needs. The main elements of an Intrusion Detection System include: a) the Intrusion Detection Sensor(s), b) the Alarm Processor, c) the Intrusion/Alarm Monitoring Station, and d) the communications structure that connects these elements and connects the system to the reaction elements. However, all systems also include people and procedures, both of which are of equal and possibly greater importance than the individual technology aspects of the system. In order to effectively utilize an installed security system, personnel are required to operate, monitor and maintain the system, while an equally professional team is needed to assess and respond to possible intrusions.

Intrusion detection sensors discussed in this Handbook have been designed to provide perimeter security and include sensors for use in the ground, open areas, inside rooms and buildings, doors and windows. They can be used as stand alone devices or in conjunction with other sensors to enhance the probability of detection. In the majority of applications, intrusion detection sensors are used in conjunction with a set of physical barriers and personnel/vehicles access control systems. Determining which sensor(s) are to be employed begins with a determination of what has to be protected, its current vulnerabilities, and the potential threat. All of these factors are elements of a Risk Assessment, which is the first set in the design process.

1.8   PERFORMANCE CHARACTERISTICS:

In the process of evaluating individual intrusion detection sensors, there are at least three performance characteristics which should be considered: Probability of Detection (PD), False Alarm Rate (FAR), and Vulnerability to Defeat (i.e. typical measures used to defeat or circumvent the sensor).

A major goal of the security planner is to field an integrated Intrusion Detection System (IDS) which exhibits a low FAR and a high PD and is not susceptible to defeat.

Probability of Detection provides an indication of sensor performance in detecting movement within a zone covered by the sensor. Probability of detection involves not only the characteristics of the sensor, but also the environment, the method of installation and adjustment, and the assumed behavior of an intruder.

False Alarm Rate indicates the expected rate of occurrence of alarms which are not attributable to intrusion activity. For purposes of this Handbook, *"false alarms"* and *"nuisance alarms"* are included under the overall term "False Alarm Rate", although technically, there is a distinction between the two terms. A *nuisance alarm* is an alarm event which the reason is known or suspected (e.g. animal movement/electric disturbance) was probably not caused by an intruder. A *false alarm* is an alarm when the cause is unknown and an intrusion is therefore possible, but a determination after the fact indicates no intrusion was attempted. However, since the cause of most alarms (both nuisance/false) usually cannot be assessed immediately, all must be responded to as if there is a valid intrusion attempt.

Vulnerability to Defeat is another measure of the effectiveness of sensors. Since there is presently no single sensor which can reliably detect all intruders, and still have an acceptably low FAR, the potential for "defeat" can be reduced by designing sensor coverage using multiple units of the same sensor, and/or including more than one type of sensor, to provide overlapping of the coverage area and mutual protection for each sensor.

1.9   ENVIRONMENTAL CONSIDERATIONS

Most security zones have a unique set of environmental factors which are taken into consideration when designing the system, selecting the sensors, and performing the installation. Failure to consider all the factors can result in excessive "false alarms" and/or "holes" in the system.

Each potential intrusion zone, whether it be a perimeter fence, an exterior entrance, a window, an interior door, a glass partition or a secured room, will have special "environmental" factors to be considered. External zones are likely to be affected by the prevailing climate, daily/hourly fluctuations in weather conditions, or random animal activity as well as man-made "environmental" factors such as activity patterns, electrical fields and/or radio transmissions, and vehicle, truck, rail or air movement.

There are a wide variety of other considerations which must be assessed when placing sensors to monitor the perimeter of an area or building. A fundamental consideration is the need to have a well-defined clear/surveillance or isolation zone. Such a zone results in a reduction of FARs caused by innocent people, large animals, blowing debris, etc. If fences are used to delineate the clear zone or isolation zone, they should be carefully placed, well constructed and solidly anchored, since fences can move in the wind and cause alarms. Consideration should also be given to dividing the perimeter into independently alarmed segments in order to localize the area of the possible intrusion and improve response force operations.

Internal zone sensors can also be impacted by a combination of external stimuli, such as machinery noise and/or vibrations, air movement caused by fans or air conditioning/heating units, and changes in temperature to mention a few. Many of these and others will be discussed in the individual Technology Reviews presented in Section Two.

1.10   ALARM MONITORING SYSTEMS

In addition to the Off-the-Shelf Intrusion Technology that is discussed in this Handbook, there is a variety of alarm monitoring systems available. Although each system is unique in the number and variety of options available, all systems perform the basic function of annunciating alarms and displaying the alarm locations in some format. The front-end (control function) of most of these systems is configured with standard 486 or Pentium computer utilizing Windows, DOS, UNIX or OS/2 as the operating system. Many of these systems operate with proprietary software, written by the manufacturer of the security system.

1.11   ALARM ASSESSMENT

State-of-the-art alarm assessment systems provide a visual and an audible indication of an alarm. The alarm data is displayed in one of two forms - either as text on a computer/monitor screen or as symbols on a map representation of the area. Most systems offer multiple levels (scales) of maps which can be helpful in guiding security personnel to the location of the alarm. The urgency of the audible/visual alarm cue can vary as to the nature of the alarm or the location of the possible intrusion (e.g. high priority versus low priority areas). In most security systems, several of these capabilities are combined to provide the Security Operations Center personnel with a relatively comprehensive picture of the alarm situation. One option offers a visual surveillance capability which automatically provides the Security Alarm Monitor with a real-time view of the alarm/intrusion zone.

1.12 SENSOR INTEGRATION

From a technology perspective, the integration of sensors into a coherent security system has become relatively easy. Typically, most sensor systems have an alarm relay, from points a, b or c, and may have an additional relay to indicate a tamper condition. This relay is connected to field panels via four wires, two for the alarm relay and two for the tamper relay, or two wires, with a resistive network installed to differentiate between an alarm and tamper condition. Most monitoring systems will also provide a means of monitoring the status of the wiring to each device. This is called line supervision. This monitoring of the wiring provides the user with additional security by indicating if circuits have been cut or bypassed.

Additionally, different sensors can be integrated to reduce false alarm rates, and/or increase the probability of intrusion detection. Sensor alarm and tamper circuits can be joined together by installing a logic "and" circuit. This "and" system then requires multiple sensors to indicate an alarm condition prior to the field unit sending an alarm indication. Usage of the logic "and" circuit can reduce false alarm rates but it may decrease the probability of detection because two or more sensors are required to detect an alarm condition prior to initiating an alarm .

1.13 COMMUNICATIONS

Communications between the front-end computer and the field elements (sensors, processors) usually employ a variety of standard communications protocols. RS-485, RS-232, Frequency Shift Keying (FSK), and Dual Tone Multi Frequency (DTMF) dial are the most common, although occasionally manufacturers will use their own proprietary communications protocol which can limit the option for future upgrades and additions. In order to reduce the tasks required to be handled by the computer, some systems require
a preprocessing unit located between the computer and the field processing elements. This preprocessor acts as the communications coordinator to "talk" to the field elements thus relieving the computer of these responsibilities.

1.14 POWER SUPPLY

Regardless of how well designed and installed, all intrusion detection systems are vulnerable to power losses, and many do not have an automatic restart capability without human intervention. Potential intruders are aware of this vulnerability and may seek to "cut" power if they cannot circumvent the system via other means. It is critical that all elements of the system have power backups incorporated into the design and operation to guarantee uninterrupted integrity of the sensor field, alarm reporting, situation assessment, and response force reaction.

1.15   COST CONSIDERATIONS

The costs of an Intrusion Detection System are easy to underestimate. Sensor manufacturers often quote a cost per meter, cost per protected volume, for the sensor system. Often this figure is representative of the hardware cost only, and does not include the costs of installation, any associated construction or maintenance. Normally, the costs associated with procuring the sensor components are outweighed by the costs associated with acquiring and installing the assessment and alarm reporting systems.

1.16   SENSOR APPLICATIONS

Most sensors have been designed with a specific application in mind. These applications are categorized by the environment where they are most commonly employed. The two basic environments or categories are Exterior and Interior. Each of the two basic categories has a number of sub-sets, such as fence, door, window, hallway, and room.

The first two of the following set of graphics show a "family tree" illustration of the sensors most applicable to these two environments (exterior/interior). As mentioned previously, some of the technologies can be used in both environments, and consequently are shown on both graphics.

SUPPLEMENTAL GRAPHIC REPRESENTATIONS

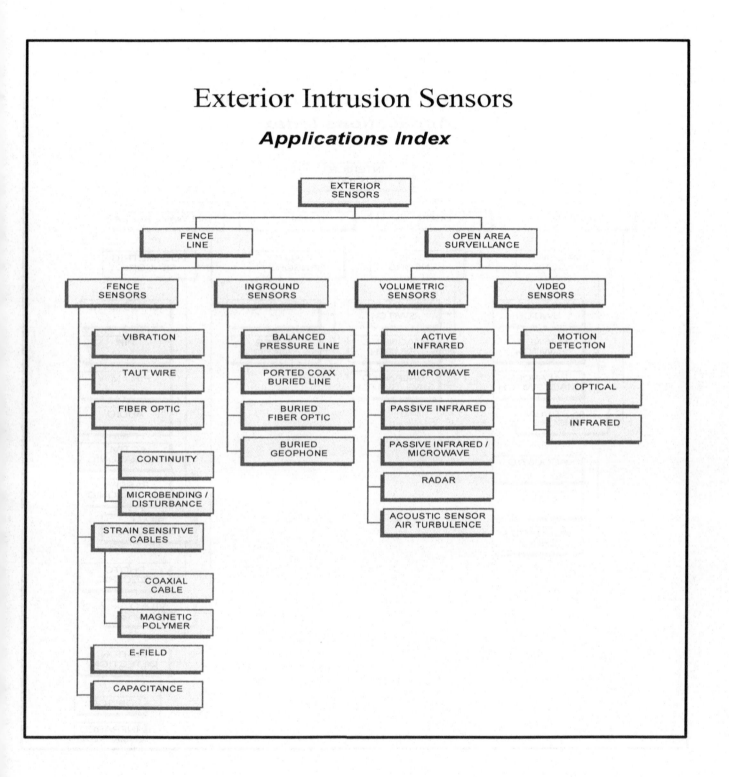

Exterior Intrusion Sensors

Applications Index

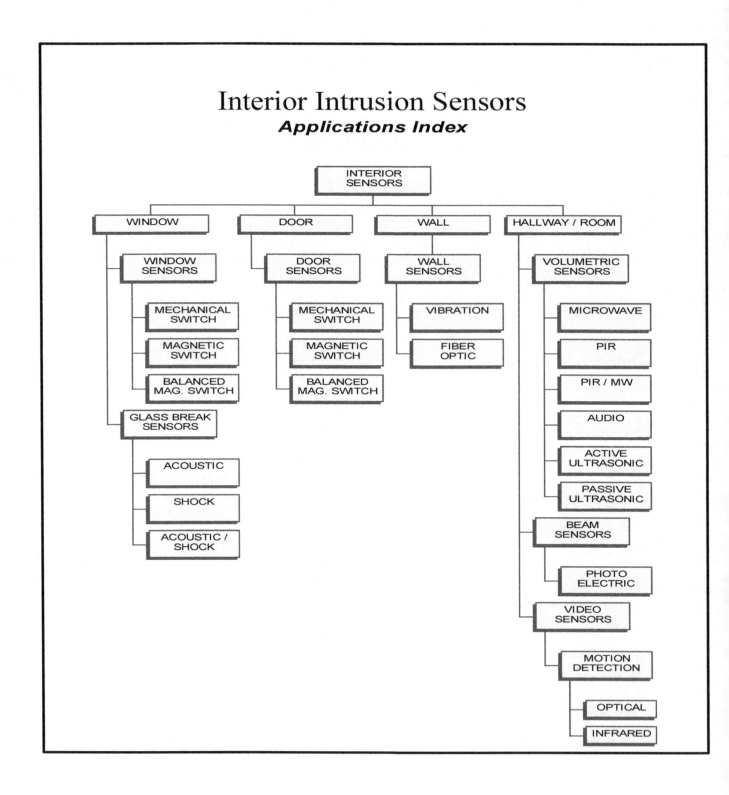

Interior Intrusion Sensors
Applications Index

TYPICAL PERIMETER SECURITY INTRUSION DETECTION PROCESS

(A) THE INTRUSION SENSOR DETECTS INTRUSION, CAUSED BY CROSSING OF THE PERIMETER ZONE

(B) THE SIGNAL IS ANALYZED BY THE SENSOR PROCESSOR TO DIFFERENTIATE BETWEEN LEGITIMATE AND UNWARRANTED CHARACTERISTICS RELATING TO INTRUSION

(C) ONCE THE SIGNAL IS DETERMINED TO BE CHARACTERISTIC OF AN INTRUSION ATTEMPT AN ALARM IS GENERATED OR THE COMMAND POST IS ALERTED

(D) AFTER THE COMMAND POST IS ALERTED, THE RESPONSE FORCE IS NOTIFIED FOR ASSESSMENT

EXTERIOR SENSOR APPLICATIONS MODEL

This example is typical of a secured facility employing various detection sensors.
The illustration shows how sensors can operate in conjunction with each other, and hypothetically
where sensors would be installed to enhance intrusion detection probability.

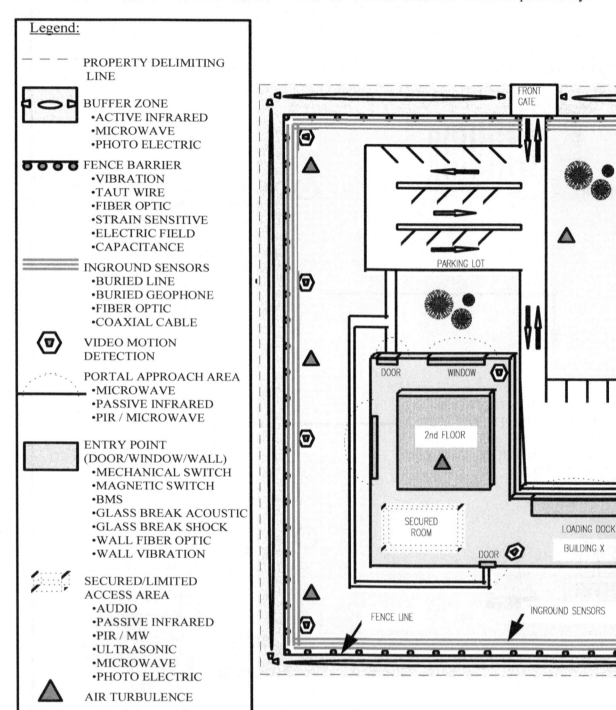

Legend:

– – – PROPERTY DELIMITING LINE

BUFFER ZONE
- ACTIVE INFRARED
- MICROWAVE
- PHOTO ELECTRIC

FENCE BARRIER
- VIBRATION
- TAUT WIRE
- FIBER OPTIC
- STRAIN SENSITIVE
- ELECTRIC FIELD
- CAPACITANCE

INGROUND SENSORS
- BURIED LINE
- BURIED GEOPHONE
- FIBER OPTIC
- COAXIAL CABLE

VIDEO MOTION DETECTION

PORTAL APPROACH AREA
- MICROWAVE
- PASSIVE INFRARED
- PIR / MICROWAVE

ENTRY POINT
(DOOR/WINDOW/WALL)
- MECHANICAL SWITCH
- MAGNETIC SWITCH
- BMS
- GLASS BREAK ACOUSTIC
- GLASS BREAK SHOCK
- WALL FIBER OPTIC
- WALL VIBRATION

SECURED/LIMITED ACCESS AREA
- AUDIO
- PASSIVE INFRARED
- PIR / MW
- ULTRASONIC
- MICROWAVE
- PHOTO ELECTRIC

AIR TURBULENCE

INTERIOR SENSORS APPLICATIONS MODEL

The following example is a typical secured room employing various detection sensors.
This illustration demonstrates how sensors operate in conjunction with each other, and where
sensors can be installed to enhance security.

Legend:

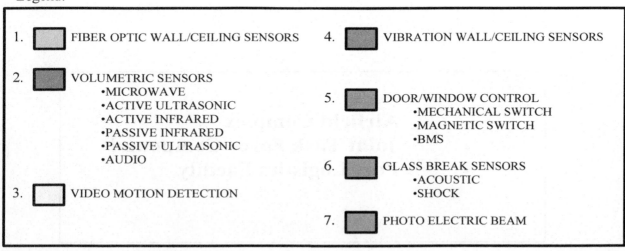

1. ▢ FIBER OPTIC WALL/CEILING SENSORS

2. ▢ VOLUMETRIC SENSORS
   - MICROWAVE
   - ACTIVE ULTRASONIC
   - ACTIVE INFRARED
   - PASSIVE INFRARED
   - PASSIVE ULTRASONIC
   - AUDIO

3. ▢ VIDEO MOTION DETECTION

4. ▢ VIBRATION WALL/CEILING SENSORS

5. ▢ DOOR/WINDOW CONTROL
   - MECHANICAL SWITCH
   - MAGNETIC SWITCH
   - BMS

6. ▢ GLASS BREAK SENSORS
   - ACOUSTIC
   - SHOCK

7. ▢ PHOTO ELECTRIC BEAM

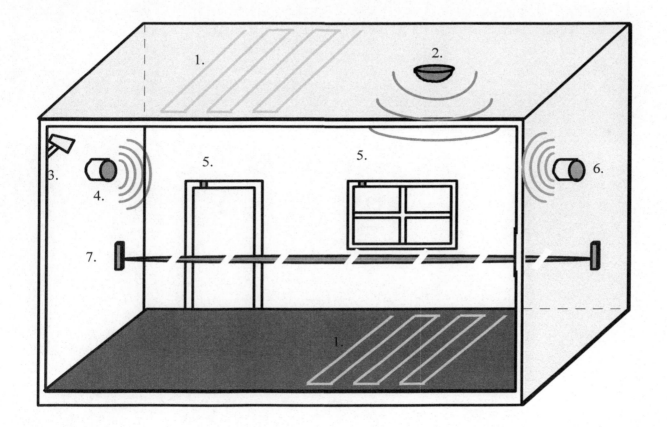

Military Application Models

- **Airfield Complex**
- **Joint Task Force Compound**
- **Port/Logistics Facility**

Airfield Complex

— · — · — · — AREA OF U.S. INTEREST (RESPONSIBILITY OF HOST NATION)
× × × × × × AREA OF RESPONSIBILITY PERIMETER

HOST NATION BUILDINGS/FACILITIES =

AREA OF U.S. INTEREST (RESPONSIBILITY OF HOST COUNTRY)

U.S. RESPONSIBILITY

U.S. BARRACKS

OPERATIONS &
MAINTENANCE

TERMINAL

RUNWAY

WAREHOUSE & TRANSSHIPMENT

FUEL

PARKING AREA

WAREHOUSE

Joint Task Force Compound

– – – – – – – AREA OF U.S. INTEREST (RESPONSIBILITY OF HOST NATION)
✗ ✗ ✗ ✗ ✗ ✗ AREA OF RESPONSIBILITY PERIMETER

HOST NATION BUILDINGS/FACILITIES =

AREA OF U.S. INTEREST (RESPONSIBILITY OF HOST NATION)

U.S. RESPONSIBILITY

COMMUNICATIONS CENTER

BARRACKS

BARRACKS

LOCAL COMMUNITY/BUSINESSES

LOCAL COMMUNITY/BUSINESSES

HEADQUARTERS COMPLEX

ARMS BLDG

GUARD

SHOP

SHOP

TRAINING & DINING COMPLEX

Port/Logistic Facility

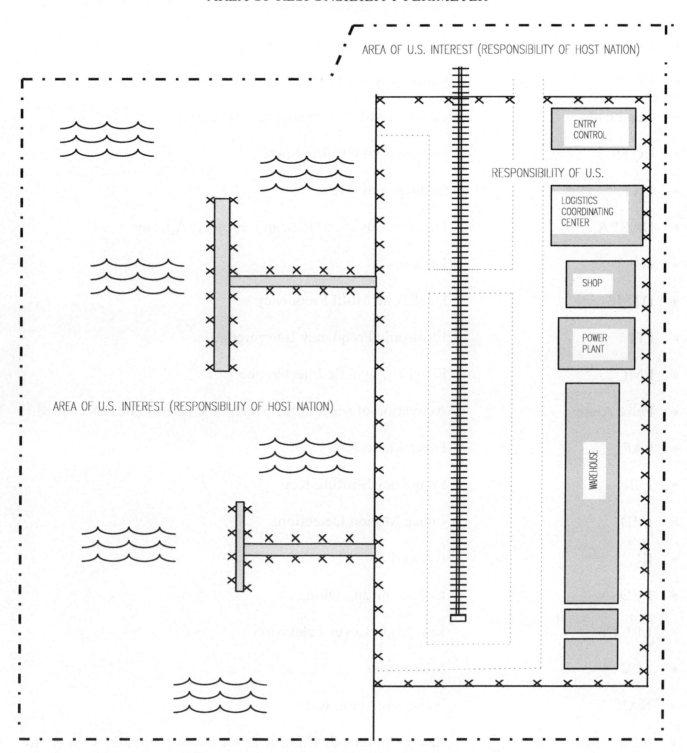

ACRONYMS AND KEY TERMS

- BMS   Balanced Magnetic Switch

- CBD   Commerce Business Daily

- CCTV   Closed Circuit Television

- Conductor   Material which transmits electric current.

- COTS   Commercial Off The Shelf

- CRT   Cathode Ray Tube

- DARPA   Defense Advanced Research Projects Agency

- dB   Decibels

- DTMF   Dual Tone Multi Frequency

- EFI   Electronic Frequency Interference

- EMI   Electro-Magnetic Interference

- False Alarm   Activation of sensors for which no cause can be determined.

- FAR   False Alarm Rate

- FSK   Frequency Shifting Key

- IMD   Image Motion Detection

- IR   Infrared

- LED   Light Emitting Diode

- LLLTV   Low Light Level Television

- MW   Microwave

- NAR   Nuisance Alarm Rate

- NISE - East   Naval Command, Control & Ocean Surveillance Center, In-Service Engineering, East Coast Division

- PD   Probability of Detection

- PIR   Passive InfRared

- RADAR   RAdio Detection And Ranging

- RCO   Receiver Cut-Off

- RF   Radio Frequency

- RFI   Radio Frequency Interference

- Rx   Receiver

- TV   Television

- Tx   Transmitter

- VMD   Video Motion Detection

KEY TERMS

- Applications

  The installation and working environments (e.g. exterior, interior, hallways, rooms), and the zone/coverage pattern that are applicable to a particular sensor. Other sensors which can provide complimentary coverage are also cited.

- Capacitance

  The property of two or more objects which enables them to store electrical energy in an electrostatic field between them.

- Causes for Nuisance Alarms

  Activity/events in which a properly operating sensor generates an alarm not attributable to intentional intrusion activity, are discussed. These activity/events are typically caused by "predictable/known" changes in the environmental "norm", such as vegetation movement, strong/turbulent weather conditions, and animal activity.

- Conditions for Unreliable Detection

  Conditions which can lower the probability of detection and effect the ability of the sensor to fully function. These conditions typically include factors such as weather, background noise, electronic interference, poor surveillance environment, obstructions, and indiscriminate placement of "foreign" objects (e.g. boxes, vehicles).

- Typical Defeat Measures

  Typical methods which may be used by an intruder to bypass/avoid detection.

SECTION TWO

SENSOR TECHNOLOGY REVIEWS

This section presents information on twenty-eight Intrusion Detection Sensor Technologies. Each sensor technology is discussed separately and has been sequenced to move from the more familiar to the more complex. The reviews have also been grouped to flow from interior point security systems to wall systems, to controlled area coverage systems, to exterior perimeter systems (including a variety of "fence" systems), and then to buried "cordon violation" systems. The last several categories, Image (Video) Motion Detection, Radar and Acoustic Air Turbulence represent newer capabilities.

Additional information, in the form of drawings, is located at the end of each Review. Although there are some minor differences in sub-paragraphing in a few of the technologies, the overall framework and key paragraph headings and content are consistent.

The basic format is as follows:

1.	Introduction	
2.	Operating Principles	
3.	Sensor Types/Configurations	
	a.	Type One
	b.	Type Two (if applicable)
4.	Applications and Considerations	
	a.	Applications
	b.	Conditions for Unreliable Detection
	c.	Causes for Nuisance Alarms
5.	Typical Defeat Measures	

SECTION TWO   TECHNOLOGY REVIEWS

TECHNOLOGY REVIEW # 1

MECHANICAL SWITCH

1. Introduction: Mechanical switches are used to detect the opening of a protected door or window. These sensors are *contact switches* that depend on direct physical operation/disturbance of the sensor to generate an alarm.

2. Operating Principle: Mechanical switches are spring-loaded or plunger devices that trigger when a door or window is opened.

3. Applications and Considerations:

    a. Applications: Mechanical switches can be mounted on doors, windows, drawers, cabinets to detect opening. They are best used in conjunction with a motion detector device, located inside the room/container, in case intrusion is made by bypassing the switch. To be effective, doors and windows should be properly and securely seated/mounted in their supporting frame prior to the installation of any security (or locking) devices including mechanical switches.

    b. Conditions for Unreliable Detection: Poor/lose fitting doors or windows can create conditions for unreliable detection, as lose mounting will allow random movement of a door or window to trigger an alarm and could assist a knowledgeable intruder in gaining surreptitious entry.

    c. Major Causes for Nuisance Alarms: Poor fitting doors or windows. Improper installation of doors, windows, locks or alarm switches are the primary cause of NAR. In addition, alarms caused by lose fitting or improperly mounted doors or windows can be aggravated by extreme weather conditions (wind and storms) as well as seasonal fluctuations in the external and/or internal environment (heating versus air conditioning).

4. Typical Defeat Measures: Holding the switch in the "normal closed" position while opening the door or window will preclude the initiation of an alarm. Typically this is accomplished with a small piece of metal designed to prevent the switch from triggering. Also, taping the switch in the "closed" position during daytime operations allows an intruder to return after the alarm has been activated and open the door or window without generating an alarm.

TECHNOLOGY REVIEW # 2

MAGNETIC SWITCH

1.   Introduction: Magnetic switches are *contact switches* used to detect the opening of a door or window and depend on the direct physical operation/disturbance of the sensor to generate an alarm.

2.   Operating Principle: Magnetic switches are composed of two parts - a two-position magnetic switch mounted on the interior of a door, window or container frame, and a two-position, magnetically operated switch. The standard switch is designed to be either normally open or normally closed, depending on the design. When the door/ window is closed, the magnet pulls the switch to its "normal" non-alarmed position. When the door/ window is opened, the magnet releases the switch, breaking the contact and activating the alarm.

3.   Applications and Considerations:

  a.   Applications: Magnetic switches are mounted on doors, windows and containers to detect opening. In high value circumstances, they should be used in conjunction with a motion detector sensor located inside the room to detect an intrusion made other than via the alarmed door, window or access portal.

  b.   Conditions for Unreliable Detection: Excessive movement of the door, window or access panel in its frame/setting can generate conditions for unreliable detection and should be corrected prior to installation of the security switches.

  c.   Major Causes for Nuisance Alarms: Poor fitting doors or windows (caused by age or and improper installation) and compounded by extreme weather conditions which cause excessive movement of the door or window are the major causes of nuisance alarms.

4.   Typical Defeat Measures: Penetration of the door or window without moving the magnet switch mechanism will bypass the alarm device. A second, free-moving and stronger magnet can be used to imitate the mounted magnet, allowing the door to be opened without generating an alarm. The location of the switch should not be observable to a potential intruder, reducing an intruder's ability to bypass or "jump" the terminal.

TECHNOLOGY REVIEW # 3

BALANCED MAGNETIC SWITCH (BMS)

1.   **Introduction:** Balanced Magnetic Switches consist of a *switch assembly* with an internal magnet that is usually mounted on the door/window frame and a *balancing (or external) magnet* mounted on the moveable door/window.

2.   **Operating Principle:** Typically, the switch is balanced in the open position between the magnetic field of the two magnets. If the magnetic field is disturbed by the movement of the external magnet, the switch moves to a "closed" position. When the door is in the normal closed position, the magnetic field generated by the biasing magnet interacts with the field created by the switch magnet, so that the total net effect on the switch is stable. When the door is opened, the switch falls to one of the contacts, becoming unstable and generating an alarm.

3.   **Applications and Considerations:**

    a.   **Applications:** Balanced Magnetic Switches (BMS) provide a higher level of security for windows and doors than magnetic or mechanical switches. Balanced magnetic switches are available in casings designed to prevent the switch from electrically causing an explosion in a hazardous area. These switches are recommended for flammable or hazardous environments. The balanced magnetic switch should be mounted on the door frame, and the balancing magnet on the door. The switch should be adjusted to initiate an alarm when the door is opened between a half and one inch. For enhanced security, a BMS (just as mechanical and straight magnetic switches) should be used in conjunction with a motion detector located inside the room, hallway or container in case intrusion is made by bypassing the switch.

    b.   **Conditions for Unreliable Detection:** Excessive movement in the door or window will create conditions for unreliable detection and should be eliminated before security switches are installed.

    c.   **Major Causes for Nuisance Alarms:** Poorly fit doors or windows and improper installation are the primary causes of nuisance alarms. Extreme weather conditions which cause excessive movement of the door, window or access portal can add to the NAR.

4.   **Typical Defeat Measures:** A distinct advantage to using the balanced magnetic switch is its inherent ability to counter a common defeat measure used on straight magnetic sensors. This defeat measure involves placing an external magnet on the switch housing to hold the internal switch in place while the door or window is opened. The design of the Balance Magnetic Switch precludes this defeat mechanism from being effective.

BALANCED MAGNETIC SWITCH

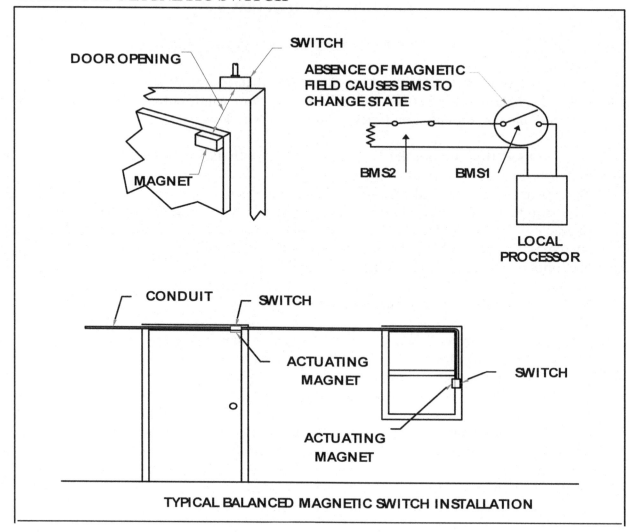

TYPICAL BALANCED MAGNETIC SWITCH INSTALLATION

TECHNOLOGY REVIEW # 4

GLASSBREAK

1.   Introduction: Glassbreak sensors monitor glass that is likely to be broken during intrusion.   The sensors are housed in a single unit and mounted on a stable interior element (wall or ceiling) facing the main glass surface. Three types of sensors are used: acoustic, shock, and a dual technology (shock/acoustic) sensor.   Regardless of which sensor is used, coverage typically does not exceed 100 square feet of glass surface.

2.   Operating Principle: Glassbreak sensors use a microphone to listen for frequencies associated with breaking glass.   A processor filters out all unwanted frequencies and only allows the frequencies at certain ranges to be analyzed.   The processor compares the frequency received to those registered as being associated with the breaking of glass.   If the received signal matches frequencies characteristic of breaking glass, then an alarm is generated.

3.   Sensor Types/Configurations: There are three basic types of glass break sensors - acoustic, shock, and a combination of the two, resulting in a dual technology (acoustic / shock) sensor.

     a.   Acoustic Sensors: Acoustic sensors listen for, and detect, the high frequency typically created when an initial shattering impact is made on the window. Once impact is made, high frequencies caused by the glass breaking travel away from the point of impact toward the outer edges of the glass surface. These vibrations excite the acoustic sensor processor which passes the frequency through a filter, compares the frequency for a match and signals an alarm if appropriate.

     b.   Shock Sensors: Shock sensors feel/sense the typical 5 KHz frequency shock wave that is created when glass is broken. When the processor detects this shock it signals an alarm. Two types of "shock" sensors (transducers) are used: *electric piezo* and *non-electric piezo*. Most use piezo transducers to "feel/sense" the 5 KHz frequency. However, some use a non-electric piezo transducer which does not have any electricity present until the piezo "bends" when it is "hit" by a 5 KHz. signal. The non-electric piezo type reduces false alarms dramatically.

     c.   Dual Technology Acoustic/Shock Sensors: In dual-tech sensors an acoustic device is linked with a shock device. This combination   utilizes the complementary capabilities of both devices and provides for a low false alarm rate sensor. The two sensing elements are located within a single casing unit, and are connected electronically through the use of an *AND* logic function.

The acoustic portion of the sensor uses a microphone to detect frequencies associated with breaking glass. A processor filters out all unwanted frequencies and only allows frequencies at certain ranges to be analyzed. Once the processor receives the frequency, it is compared to those associated with glass breakage. If the signal matches frequencies characteristic of breaking glass, then a signal is sent to the *AND* gate.

The shock portion of the sensor "feels" for the 5 KHz frequency in the form of a shock wave created when glass is broken. When the processor detects this shock, it sends a signal to the *AND* gate. Once the *AND* gate has received both signals an alarm is generated.

NOTE: A distinct advantage to this sensor is its incorporation of two Glassbreak technologies into one sensor. This significantly reduces false alarms from background noise such as RFI and frequency noise created by office machines.

4. Applications and Considerations:

a. Applications: Depending on the manufacturer's specifications, acoustic sensors should be mounted on the window, window frame, wall or ceiling. If mounted on the glass, the sensor should be placed in the corner approximately two inches from the edge of the frame. If mounted on the wall or ceiling, the sensor should be installed opposite the window.

Glassbreak sensors should be used in conjunction with contact switches (e.g., magnetic switches, balanced magnetic switches) in case intrusion is attempted by opening the window instead of breaking it.

A volumetric (area monitoring) motion detector should also be incorporated in the protected interior area to detect intrusion/entry by an avenue other than the window. The volumetric device should be positioned at a point and angle that allows it to look in toward the window of concern to maximize the detection capability.

NOTE: Although not recommended, the sensor may be mounted on the window. If so, the mounting adhesive should be specified to withstand long exposure to summer heat, winter cold and condensation that might collect on the window. It should be noted that a window glass can get as hot as 150° F in the summer and as cold as -30° F in the winter, therefore, it is essential that the application adhesive meets these specifications.

b. Conditions for Unreliable Detection: Although inappropriate matching of sensor range capacity to the window size and poor location may cause the sensor to be out of effective detection range, the most typical deficiency occurs when the acoustical characteristics of the room are in conflict with the sensor's performance specifications. "Soft" acoustic rooms (e.g. carpeted with window drapery) that absorb vibration or by altering the acoustic characteristics of the "hard" room (e.g., adding window shutters, blinds, draperies, rugs) after the sensor has been tuned can cause detection inadequacy of the sensor.

NOTE: As a precaution all windows should be checked for cracks and replaced prior to installation of a Glassbreak sensor to ensure that a good frequency signature will be produced if the window is broken.

      c.   **Causes for Nuisance Alarms:** Improper calibration or installation of an acoustic Glassbreak sensor will cause nuisance alarms. In addition, RF interference and sharp impact noises can cause false alarms. Also, improper application/placement of the sensor or background noise, such as office, industrial and cleaning machinery, can create noise in the frequency detection range of the sensor.

5.   **Typical Defeat Measures:** The detaching/cutting of an opening in the window or the removal of a window pane (with or without a sensor mounted on it) can bypass the sensor. The break frequency can be distorted by muffling the sound of the breaking glass reducing the potential for the "correct" frequency registered by the sensor.

GLASS-BREAK SENSOR

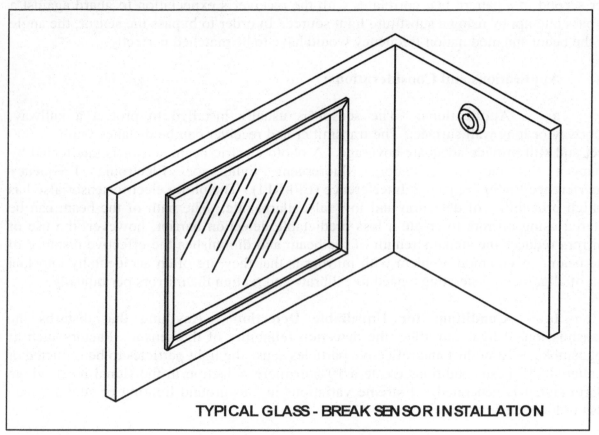

TYPICAL GLASS - BREAK SENSOR INSTALLATION

TECHNOLOGY REVIEW # 5

PHOTO ELECTRIC BEAM

1.   Introduction: Photo electric beam sensors transmit a beam of infrared light to a remote receiver creating an "electronic fence". These sensors are often used to "cover" openings such as doorways or hallways, acting essentially as a trip wire. Once the beam is broken/interrupted, an alarm signal is generated.

2.   Operating Principle: Photoelectric beam sensors consist of two components: a transmitter and a receiver. The transmitter uses a Light Emitting Diode (LED) as a light source and transmits a consistent infrared beam of light to a receiver. The receiver consists of a photoelectric cell that detects when the beam is present. If the photo electric cell fails to receive at least 90% of the transmitted signal for as brief as 75 milliseconds (time of an intruder crossing the beam), an alarm signal is generated.

The beam is modulated at a very high frequency which changes up to 1,000 times per second in a pattern that correlates with the receiver's expectation to guard against a bypass attempt by using a substitute light source. In order to bypass the sensor, the angle of the beam and modulation frequency would have to be matched perfectly.

3.   Applications and Considerations:

        a.   Applications: The sensor is usually installed to protect a hallway, doorway or long wall surface. The transmitter and receiver can be distanced up to 1,000 feet and still provide adequate coverage. A photo electric beam sensor is unaffected by changes in thermal radiation, fluorescent lights or Electronic Frequency Interference/Radio Frequency Interference (EFI/RFI). The photo electric sensor also has a high probability of detection and low false alarm rate. The path of the beam can be altered using mirrors to create a less predictable detection barrier, however, the use of mirrors reduces the signal strength of the beam and diminishes the effective distance of the beam. A common problem with mirrors is that they are often accidentally knocked out of alignment, generating a need to calibrate and realign the mirrors periodically.

        b.   Conditions for Unreliable Detection: Anything that disturbs the transmission of light can affect the detection reliability of the sensor. Factors such as fog, smoke, mist or dust and reflective particles cause the light particles to be refracted or scattered. If these conditions create a 10% or more reduction in the signal received, an alarm signal is generated. Extreme variations in background lighting or sunlight may also reduce sensitivity.

        c.   Causes for Nuisance Alarms: Any objects that may break the beam such as birds, animals, blowing leaves or paper will interrupt the signal, therefore generating an alarm. In addition, improper alignment of the transmitter, receiver or mirrors may generate an alarm. Mirrors can also collect dust, causing refraction/diffusion of the reflected beam.

4. **Typical Defeat Measures:** Stepping over or passing under the signal path will defeat the intent of the sensor. However, mirrors can be used to counter this vulnerability by creating a "Zig-Zag" multiple beam barrier pattern.

TECHNOLOGY REVIEW # 6

MICROWAVE SENSORS

1.   Introduction: Microwave sensors are motion detection devices that transmit/flood a designated area/zone with an electronic field. A movement in the zone disturbs the field and sets off an alarm. Microwave Sensors may be used in exterior and interior applications.

2.   Operating Principle: Microwave sensors transmit microwave signals in the "X" band. These signals are generated by a Gunn diode operating within pre-set limits that do not affect humans or the operation of pacemakers. Although very little power is used, the system provides enough energy for a detector to project a signal up to 400 feet in an uninterrupted line of sight. The detection of intrusion is directly related to the Doppler frequency shift principle. Most sensors are tuned to measure the Doppler shift between 20 Hz and 120 Hz. These frequencies are closely related to the movements of humans. Objects that fail to produce a signal or produce a signal outside the tuned frequencies are ignored. Objects that fall within the range cause the sensor to generate an alarm signal.

3.   Sensor Types/Configurations: There are two basic types of microwave sensors: monostatic sensors, which have the transmitter and receiver encased within a single housing unit, and bistatic sensors, in which the transmitter and receiver are two separate units creating a detection zone between them. A bistatic system can cover a larger area and would typically be used if more than one sensor is required.

     a.   Monostatic Units: The transmitter and receiver are contained in a single dual function unit. The antenna is mounted within the microwave cavity and can be configured/shaped to cover a specific area or detection zone. The shape of the detection beam can be changed to transmit a long, slender beam or a short oval one. Monostatic microwave sensors transmit signals at two different transmitting frequencies. The frequencies are rapidly turned on and off, first at one frequency and then at the other. The receiver is then shut off for a short period of time after transmission. Because microwaves travel at a constant speed and the receiver is looking for reflected energy, the receiver can be programmed to receive only signals that are able to go out and return within a specific time period. The area where all reflected frequencies can be picked up by the transmitter is known as the Receiver Cut Off (RCO) region. This enables the user to protect a well defined detection zone. The receiver is programmed to ignore signals from stationary objects and only receive signals from disturbances/movement in the designated field of coverage.

b. **Bistatic Units:** The transmitter and receiver for bistatic microwave sensors are separate units. The detection zone is created between the two units. The antenna can be configured to alter the signal field (width, height), creating different detection zones. The receiver is programmed to receive signals from the transmitter and detect a change in the frequencies caused by a movement in the field of coverage. Bistatic microwave sensors transceivers are somewhat limited by poorly defined detection patterns, and nuisance alarms may be a problem if large metal objects are nearby or if windy conditions exist.

4. **Applications and Considerations:**

a. **Applications:** Microwave sensors can be used to monitor both exterior areas and interior confined spaces, such as vaults, special storage areas, hallways and service passageways. In the exterior setting they can be used to monitor an area or a definitive perimeter line, as well as to serve as an early warning alert of intruders approaching a door or wall. In situations where a well-defined area of coverage is needed, monostatic microwave sensors should be used. However, monostatic microwave sensors are limited to 400 feet coverage, while bistatic sensors can extend up to 1,500 feet. To further enhance detection, video motion detection equipment (or another type sensor) can be installed to complement the microwave application. The use of a companion system, such as video image motion detection, not only provides a second line of defense, but provides security personnel with an additional tool to assess alarms and discriminate actual/potential penetrations from false alarms or nuisance events.

b. **Conditions for Unreliable Detection:** Since microwave sensors operate in the high frequency spectrum (X band), close association or proximity to other high frequency signals can adversely affect the detection reliability of these sensors.

Areas that contain strong emitters of electric fields (radio transmitters) or magnetic fields (large electric motors or generators) can effect the ability of microwave sensors to function properly, and should be avoided or compensated for by distinct signal separation.

Zones that contain fluorescent lights can also pose a problem. The ionization cycle created by fluorescent bulbs can be interpreted by the detector as motion and thus provide false alarms.

Self generated signal reflection is a common problem caused by improper placement/mounting. Positioning the sensor externally and parallel to the wall rather than imbedding it in the wall will avoid this problem.

Also, large metal objects which can reflect the signal and/or provide "dead pockets" should be kept out of the detection zone, as should equipment whose operation involves external movement or rotating functions.

c.   **Causes for Nuisance Alarms:** Because of the high frequencies at which microwaves travel, the signal/sensor is not affected by moving air, changes in temperature or humidity. However, the high frequency allows the signal to easily pass through standard walls, glass, sheet rock, and wood. This can cause false alarms to be generated by movement adjacent to, but outside the protected area. Conversely, it is essential to test for, note, and compensate for any dead spots (areas of no detection) created by metal objects such as dumpsters, shipping crates, trash cans, and electrical boxes. These dead spots create ideal areas for intrusion attempts. In addition, signals reflected off these type objects/materials can "extend" sensor coverage to areas not intended to be covered, thus creating the potential for false alarms.

5.   Typical Defeat Measures: An intruder with some degree of periodic access to the denied area may be in a position to conduct "walk tests" or otherwise cause/observe the alarm activation pattern, and determine nominal detection coverage patterns, thereby identifying a possible low detection approach path. In addition, an intruder advancing at a deliberately slow rate of movement, who takes maximum advantage of any obscuring, blocking or signal absorbing characteristics associated with the surveillance environment can reduce the probability of detection. However, regular calibration of the sensor(s), sanitation of the area, and the use of another type of sensor can substantially increase the probability of detection.

TYPICAL LONG RANGE DETECTION PATTERN FOR MONOSTATIC MICROWAVE SENSORS

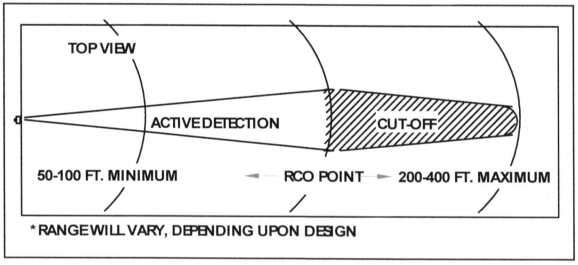

TYPICAL SHORT RANGE MONOSTATIC MICROWAVE DETECTION PATTERN

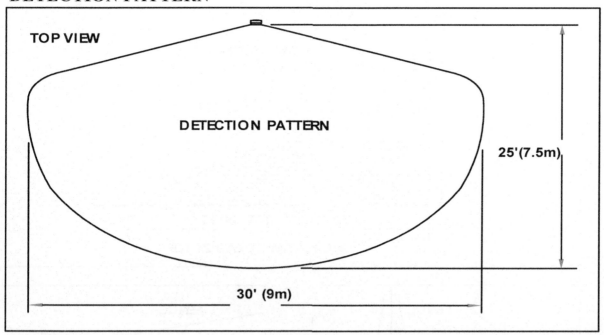

TOP VIEW

DETECTION PATTERN

25' (7.5m)

30' (9m)

TYPICAL BISTATIC MICROWAVE DETECTION PATTERN

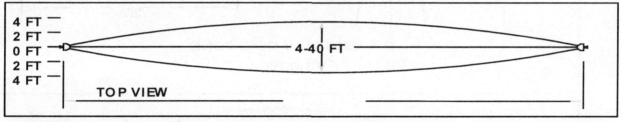

4 FT
2 FT
0 FT
2 FT
4 FT

4-40 FT

TOP VIEW

* LENGTH AND WIDTH OF DETECTION PATTERNS WILL VARY, DEPENDING UPON DESIGN.

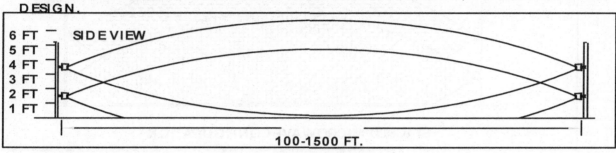

6 FT
5 FT
4 FT
3 FT
2 FT
1 FT

SIDE VIEW

100-1500 FT.

* THE MICROWAVE SENSORS CAN BE MOUNTED IN A DUAL CONFIGURATION TO PROVIDE A GREATER PROBABILITY OF DETECTION.

MICROWAVE SENSOR ZONES

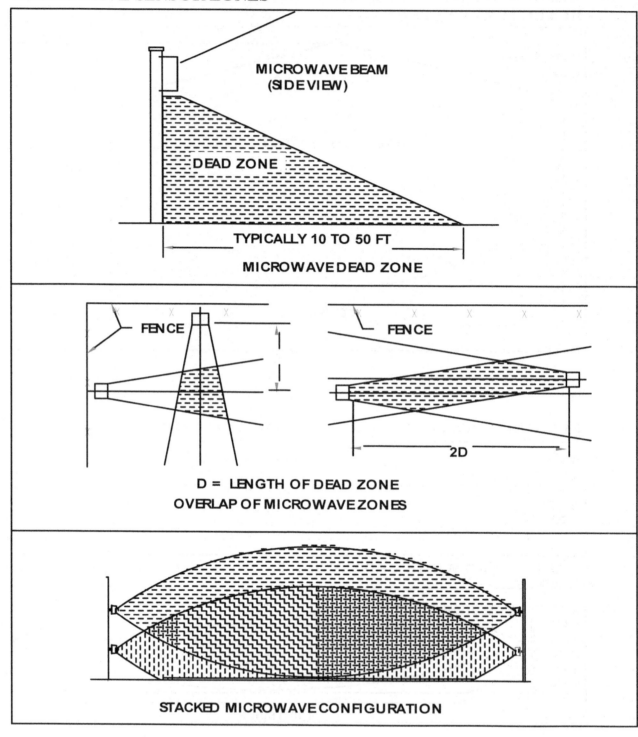

MICROWAVE BEAM
(SIDE VIEW)

DEAD ZONE

TYPICALLY 10 TO 50 FT

MICROWAVE DEAD ZONE

FENCE

FENCE

2D

D = LENGTH OF DEAD ZONE
OVERLAP OF MICROWAVE ZONES

STACKED MICROWAVE CONFIGURATION

BISTATIC MICROWAVE LAYOUT CONFIGURATIONS

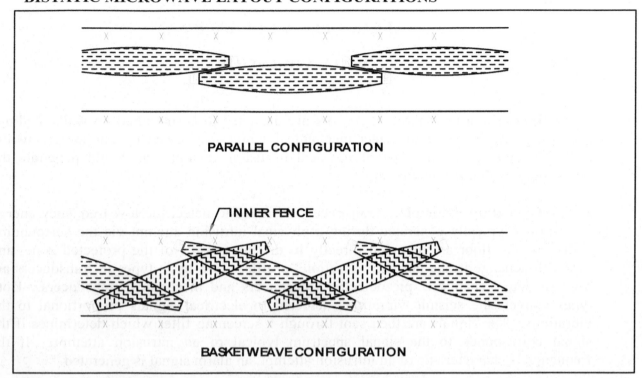

PARALLEL CONFIGURATION

INNER FENCE

BASKETWEAVE CONFIGURATION

BISTATIC MICROWAVE SENSOR

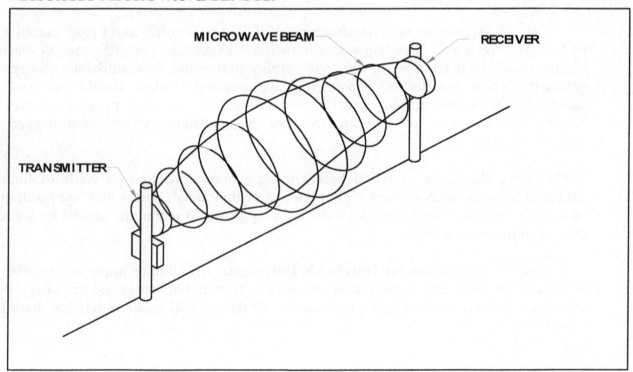

MICROWAVE BEAM

RECEIVER

TRANSMITTER

TECHNOLOGY REVIEW # 7

WALL VIBRATION

1.   Introduction: Vibration sensors are designed to be mounted on walls, ceilings and floors and intended to detect mechanical vibrations caused by chopping, sawing, drilling, ramming or any type of physical intrusion attempt that would penetrate the structure on which it is mounted.

2.   Operating Principle: Transducers designed to detect the low frequency energy (vibrations) typically generated during a physical intrusion attempt via the surrounding walls, roof or floor are mounted directly to the inner walls of the protected zone, and detect the change in the normal "vibration" profile. Two basic types of transducers are used to detect changes: piezoelectric transducers and mechanical transducers. Both types convert the seismic vibrations detected to electrical signals proportional to the vibrations. The signals are then sent through a screening filter which determines if the signal corresponds to the signal spectrum typical of an intrusion attempt. If the frequency is characteristic of an intrusion attempt, an alarm signal is generated.

3.   Applications and Considerations:

     a.   Applications: Vibration sensors should be securely and firmly placed 8 to 10 feet apart, on a wall or ceiling where intrusion is expected. The difference in spacing lengths should be determined by the wall's ability to transmit the disturbance energy. A volumetric (area monitoring) sensor (passive infrared, audio) should be used in conjunction with wall sensors and directed toward the expected penetration site, to provide detection of an intrusion that may not cause sufficient vibrations to trigger the vibration sensors.

NOTE: Care should be exercised before using vibration sensors on walls of limited structural integrity such as sheet rock, plywood or thin metal, unless they are positioned on a main support. These types of walls are very prone to vibrations caused by sources other than intrusion actions.

     b.   Conditions for Unreliable Detection: Unstable or improper installation or spacing of units, and mounting of the sensors to materials (rugs, fabric, heavy wall coverings) that are not conducive to detecting vibrations will create unreliable detection conditions.

c.   **Causes for Nuisance Alarms:**   Poor placement is a primary cause of nuisance alarms.   Vibration sensors may generate alarms if mounted on walls that are exposed to external vibrations (e.g., trains, planes), or if the walls are subject to vibrating machinery.   In any of these or similar situations, vibration sensors should not be used.

4.   **Typical Defeat Measures:** The system can be defeated by avoiding entry through the protected area, or by selecting a point and method of entry in a segment of a wall, roof or floor that will permit the suppression/diffusion of the intrusion vibrations. Another defeat measure, which is also applicable to many other sensors as well, is the generation of a persistent but random number of false alarms over a long period of time, causing the alarm to be ignored or the response time greatly diminished.

WALL VIBRATION SENSOR

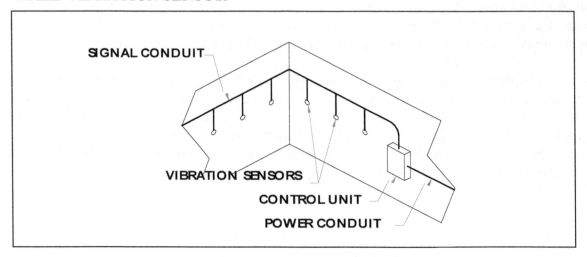

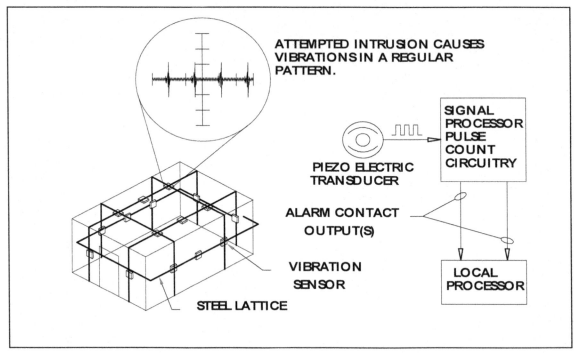

TECHNOLOGY REVIEW # 8

FIBER OPTIC WALL

1.   Introduction: A fiber optic wire sensor is in an open mesh network (quilt) appliqué that can be applied directly to an existing wall or roof, or installed in a wall (or roof) as it is being constructed. The fiber optic network is designed to detect the low frequency energy (vibrations) caused by chopping, sawing, drilling, ramming or physical attempt to penetrate the structure on which it was mounted.

2.   Operating Principle: The fiber optic cable acts as a line sensor and contains an electro optics unit which transmits light using a Light Emitting Diode (LED) as the light source. The light travels through the fiber optic network and is picked up by a detector, which is very sensitive to slight alterations in the transmission. When an adequate alteration in the light pattern takes place, the signal processor generates an alarm.

3.   Applications and Considerations:

   a.   Applications: These sensors are very sensitive, and special consideration must be given to determine if this type of sensor is suitable for a particular wall/roof. A vibration sensor may generate false alarms if mounted on walls that are exposed to external vibrations (vehicle, train or heavy foot movement) or if the walls are subject to vibrating machinery. However, an imbedded fiber optic sensor, although very perceptive to slight changes in the light pattern, can be calibrated easily and gauged to detect various forms of intrusion.

   b.   Conditions for Unreliable Detection: Improper installation or calibration. Caution should be exercised before using vibration sensors to protect walls of lesser structural integrity, such as sheet rock, plywood or thin metal. These walls are prone to vibrations from sources other than intrusion attempts.

   c.   Causes of Nuisance Alarms: Machinery that causes vibrations can generate false alarms and should be located away from the wall on which the fiber optic cable is mounted. Also, vibrations caused by exterior aircraft and train traffic can cause the wall/roof/building fabric to vibrate, thereby causing the vibration sensor to generate an alarm signal.

4.   Typical Defeat Measures: The system can be bypassed by avoiding entry through a protected area or targeting an insensitive location as the point of entry.

FIBER OPTIC STRUCTURAL VIBRATION SENSOR

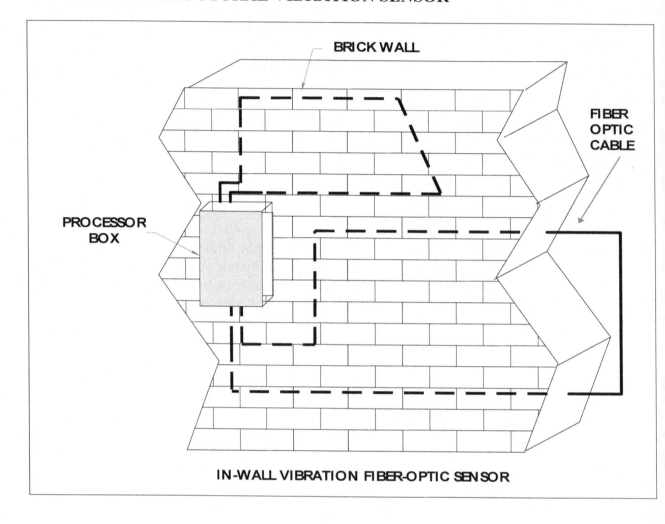

IN-WALL VIBRATION FIBER-OPTIC SENSOR

TECHNOLOGY REVIEW # 9

AUDIO SENSORS

1.   Introduction: Audio detectors listen for noises generated by an intruder's entry into a protected area, and are generally used, but not exclusively, in internal applications, from an entrance foyer to critical data/resource storage areas.

2.   Operating Principle: The sensor is made up of two devices: *Pick-up* units mounted on the walls or ceilings of the monitored area, and an *Amplifier* unit which includes processing circuitry. The Pick-up units are basically microphones that listen for noise. These microphones collect sound for analysis by the processor circuit, which can be calibrated to a noise threshold that is characteristic for an intrusion attempt. If a certain amount of noise is detected from a monitored area within a selected time period, an alarm signal is generated.

3.   Applications and Considerations:

   a.   Applications: Audio sensors should be mounted in areas where the predicted intrusion noise is expected to exceed that of the normal environmental noise. If background noise does exist, and if calibration is not accomplished to compensate for it, the microphone may be unable to detect/differentiate an intrusion noise. If excessive background noise is present, the audio sensor should not be considered.

   Typically audio sensors are used in conjunction with another detection sensor (Passive Infrared-PIR, ultra-sonic, microwave) to provide a greater probability of detection.

   Since an audio sensor is unaffected by changes in the thermal environment and fluorescent lights have no effect on the sensor's detection characteristics, its use with a thermal imaging motion detection system can provide both audio and visual record/tracking of an intrusion.

   b.   Conditions for Unreliable Detection: Principle causes of unreliable detection include ineffective sensitivity settings caused by extraneous background noise, such as clocks, office equipment, boilers and heating or air conditioning units.

   c.   Causes for Nuisance Alarms: Excessive background noise, such as airplanes, trains or loud weather (thunderstorms) may cause significant noise levels thereby generating an alarm. If these factors are present, careful consideration should be given to determining whether this sensor is appropriate.

4. **Typical Defeat Measures.** An intruder who makes a slow, deliberate entry, and takes measures to muffle the normal sounds of movement and intentionally allows sufficient lag time to occur between any noise generated by his movement may avoid detection.

TECHNOLOGY REVIEW # 10

PASSIVE ULTRASONIC

1.   **Introduction:** The passive Ultrasonic sensor is a motion detection device that "listens" for ultrasonic sound energy in a protected area, and reacts to high frequencies associated with intrusion attempts.

2.   **Operating Principle:** The *passive ultrasonic sensor* "listens" for frequencies that have a range between 20 - 30 KHz. Frequencies in this range are associated with metal striking metal, hissing of an acetylene torch, and shattering of concrete or brick. The sound generated is transmitted through the surrounding air and travels in a wave type motion. When the sound wave reaches the detection sensor it determines if the frequency is characteristic of an intrusion. If the criteria is met, an alarm signal is generated.

3.   **Applications and Considerations:**

       a.   **Applications:** Ultrasonic sensors are typically mounted on a wall or ceiling and are frequently used in tandem with another sensor, such as a passive device (Passive Infrared-PIR) to provide a greater probability of detection (P_D). However, this may also increase the overall false alarm rate (FAR) slightly, depending on the variability and uncontrollability of the environmental characteristics of the monitored area.

       An advantage to using the passive ultrasonic sensor is that the device is unaffected by heat, thus thermal changes in the environment do not hinder its detection ability. It is also easy to contain its energy within a selected area, since ultrasonic energy does not normally pass through walls, roofs or partitions. The disadvantage is that it does not pass through furniture or other obstructions either (boxes, crates), thus creating "dead zones" of non-surveillance. This disadvantage can be overcome by placing additional sensors at second and third locations to "cover" the dead zones of sensor # 1.

       b.   **Conditions for Unreliable Detection:.** Extreme changes in temperature or humidity from those prevalent during the initial installation and calibration may cause a change in detection reliability. As with most sensors, ultrasonic sensors should be recalibrated periodically, at least on a seasonal basis.

       c.   **Causes for Nuisance Alarms:** Some of the most common stimuli that cause ultrasonic sensors to alarm are air movement from heating and air conditioning systems, drafts from doors and windows, hissing from pipes, and the ringing of a telephone. All these stimuli can create noise near or in the ultrasonic range, thereby triggering an alarm.

4.   **Typical Defeat Measures:** Passive ultrasonic sensors have a limited frequency spectrum, and intrusion sounds other than those that fall into the unit's spectrum (such as drilling), will not generate an alarm signal. For this reason it is recommended that an active measures detection device (such as a microwave sensor) be used in conjunction to ensure adequate detection.

PASSIVE ULTRASONIC MOTION SENSOR

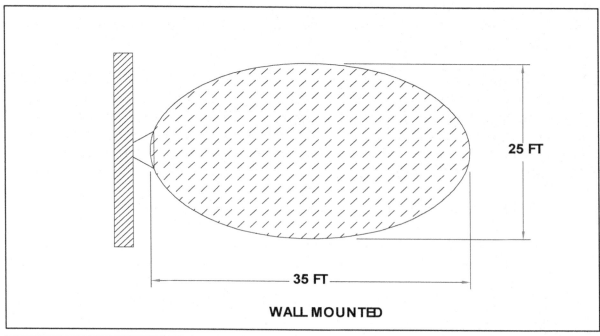

WALL MOUNTED

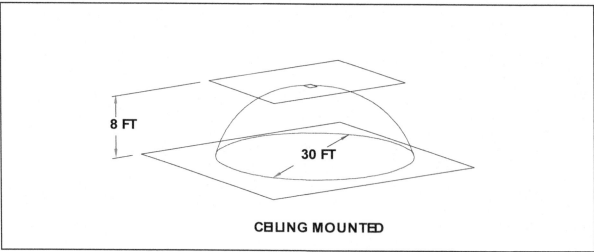

CEILING MOUNTED

TECHNOLOGY REVIEW # 11

ACTIVE ULTRASONIC

1.   Introduction: The Active Ultrasonic sensor is a motion detecting device that *emits* ultrasonic sound energy into a monitored area and reacts to a change in the reflected energy pattern.

2.   Operating Principle: Ultrasonic sensors use a technique based on a frequency shift in reflected energy to detect intruders. Ultrasonic sound is transmitted from the device in the form of energy. The sound uses air as its medium and travels in a wave type motion. The wave is reflected back from the surroundings in the room/hallway and the device "hears" a pitch characteristic of the protected environment. When an intruder enters the room, the wave pattern is disturbed and reflected back more quickly, thus increasing the pitch and signaling an alarm.

3.   Applications and Considerations:

    a.   Applications: Typically, ultrasonic sensors are mounted on the wall or ceiling. Ultrasonic sensors can be used in conjunction with a passive device (e.g., PIR) to provide a greater probability of detection (P_D). However, this may also increase the false alarm rate (FAR), depending on environmental characteristics of the monitored area.

Ultrasonic sensors are not affected by heat, thus changes in the thermal environment do not hinder its detection ability. Ultrasonic energy is easily contained within a selected area avoiding the problem of the energy passing through walls and detecting activity outside the protected zone.

    b.   Conditions for Unreliable Detection: Ultrasonic energy will not pass through most substantive objects and material, (e.g. storage, shelving), thus creating dead zones within the coverage area where the sensor is ineffective. The sensor must be positioned so dead zones are minimal. Also, extreme changes in temperature or humidity from the initial calibration may cause a hindrance in detection reliability.

    c.   Causes for Nuisance Alarms: Some of the most common stimuli that cause ultrasonic sensors to false alarm are air movement from heating, air conditioning systems, drafts from doors and windows, hissing from pipes, and telephone rings. All of these stimuli can create noise near or in the ultrasonic range, thus triggering an alarm. Also anything that causes movement, such as animals, has the potential to cause an alarm.

4.   Typical Defeat Measures:. Slow horizontal movement by an intruder across the area of coverage is often difficult for ultrasonic sensors to detect. Proper calibration is needed to ensure that slow moving intruders will be detected. In addition, a knowledgeable and properly equipped intruder can use special "test lights" to detect coverage patterns and circumvent these areas.

ACTIVE ULTRASONIC MOTION SENSOR

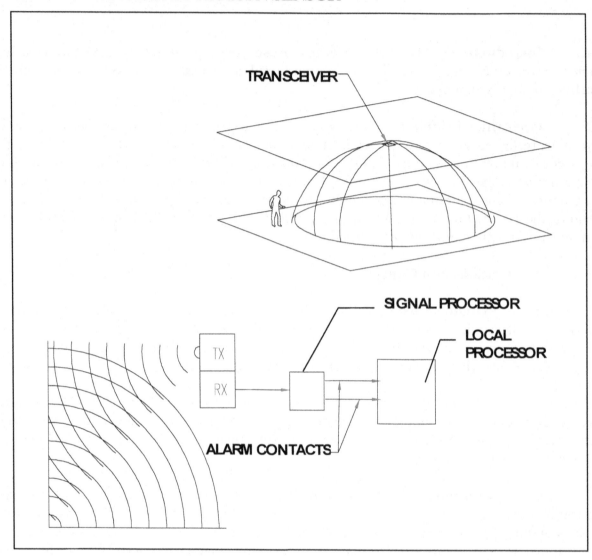

TECHNOLOGY REVIEW # 12

PASSIVE INFRARED

1.   **Introduction:** As the name implies, Passive Infrared (PIR) sensors are passive, that is, the sensor does not transmit a signal; the sensor head simply registers an impulse when received. The sensor head is typically divided into several sectors/zones, each defined with specific boundaries. Detection occurs when an emitting heat source (thermal energy) crosses two adjacent sector boundaries or crosses the same boundary twice within a specified time.

2.   **Operating Principle:** Passive infrared sensors detect electromagnetic radiated energy generated by sources that produce temperatures below that of visible light. PIR sensors do not measure the amount of IR energy per se, but rather the change of thermal radiation.   PIRs "see/detect" infrared "hot" images by sensing the contrast between the "hot" image and the "cooler" background.

Infrared energy is measured in microns, with the human body producing energy in the region of 7-14 microns. Most PIR sensors are focused on this narrow band width. In order to avoid capturing environmental thermal deviations, Rate Of Change measurement circuitry or bi-directional pulse counting circuitry is employed.

In Rate Of Change measurement, the processor evaluates the speed at which the energy in the field of view changes. Movement by an intruder in the field of view produces a very fast rate of change, while gradual temperature fluctuations produce a slow rate of change. In the bi-directional pulse counting technique, signals from separate thermal sensors produce opposite polarity. An unprotected/unshielded human entering a field of view moving at a typical speed (walk or above) will normally emit/produce several signals which allow detection to occur.

When the radiation change captured by the lens exceeds a certain pre-set value, the thermal sensor produces an electrical signal which is sent to a built-in processor for evaluation and possible alarm.

3.   **Sensor Types/Configurations:**   The PIR wavelength is subdivided into two major range detection categories: one covers Near Infrared Energy (e.g. thermal energy emitted by TV remote control devices), and the other covers the Far Infrared Energy (e.g. thermal energy emitted by people). It is this latter category which is employed in security applications.

Optics and reflective principles play a very important role in the design and function of PIRs. Because of the need to precisely focus thermal radiation, the reflection/focusing of the energy waves is done two ways: Reflective Focusing and the Fresnel Lens method.

In Reflective Focusing, the energy waves are reflected off a concave mirror and directed into the sensing element. By contrast, a Fresnel Lens allows the energy to travel directly to the sensor. Both methods use some type of protective covering on the sensor, so the loss of some energy is unavoidable. However, both sensors work quite well.

4.   Applications and Considerations:

a.   Applications: Passive infrared sensors should be installed on walls or ceilings, with the detection pattern covering the possible intrusion zones. Each detection/surveillance zone can be pictured as a "searchlight" beam that gradually widens as the zone extends farther from the sensor with different segments being illuminated while others are "dark". This design characteristic allows the user to focus the "beam" on areas where protection is needed while ignoring other areas, such as known sources of false alarms. Tower/ceiling mounted PIRs theoretically provide a 360^0 detection pattern.

b.   Curtain Lens Feature: The interchanging of different lens and reflectors/mirrors permits the field(s) of view and zones of surveillance to be changed and/or segmented. PIR design includes a "Curtain Lens" feature that provides a full barrier protection zone by eliminating the typical dead zones. PIRs with this type are ideal for protecting hallways or entry points.

c.   Conditions for Unreliable Detection: Because the PIR looks for thermal radiation projected against a cooler background, detection is based on temperature. As the environment approaches the same temperature as the intruder, the detectors become less sensitive. This is especially true for environments ranging between 80 - 100 degrees. Theoretically, if a person was radiating the same temperature as the environment, he would be invisible to the sensor. For this reason another type of sensor should be used in conjunction with the PIR to enhance the system. Complementary sensors for interior applications include balanced magnetic switches, glassbreak detectors, and time delayed CCTV cameras. For exterior applications Video Motion Detection is a good complement.

d.   Causes for Nuisance Alarms: Heat radiating from small animals and /or rodents can cause false alarms. Time activated space heaters, ovens and hot water pipes can also provide false alarms if they are in the field of view. In addition, PIR sensors that are not designed with the capacity to filter (ignore) visible light can be affected by car headlights or other sources of focused light. Although infrared energy from sunlight is filtered by ordinary window glass, objects in a room can become heated over time and subsequently begin emitting/reflecting infrared energy. If this energy is "turned off/on", (such as by the movement of clouds), it can create a random "on/off" situation, thereby generating nuisance alarms.

5. **Typical Defeat Measures:** Shadowing, cloaking or masking the intruding heat source (person/machine) from the field of view decreases the probability of detection as it reduces the possibility of sufficient radiated/emitted heat being focused on the thermal sensor. In addition, knowing the dead spots of the detection pattern can permit an intruder to bypass active regions. Walking into the sensor rather than across the sensor's field of view can also reduce the detection capability by not allowing the boundaries of the detection beams to be broken.

PASSIVE INFRARED SENSOR

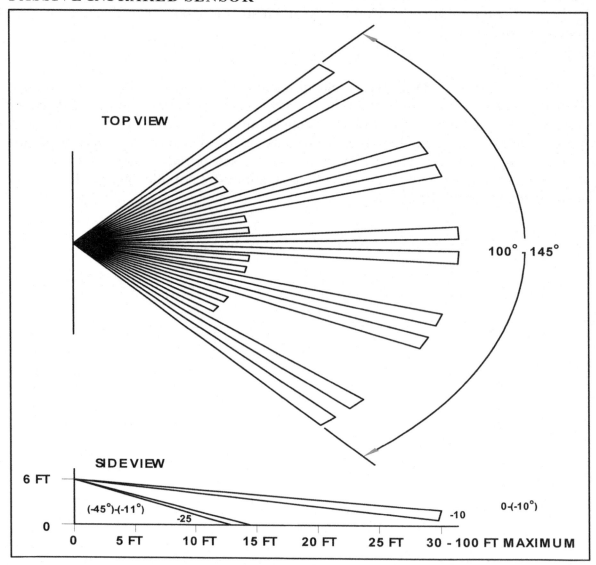

TOP VIEW

100° 145°

SIDE VIEW

6 FT

(-45°)-(-11°)

-25

0

0   5 FT   10 FT   15 FT   20 FT   25 FT   30 - 100 FT MAXIMUM

-10

0-(-10°)

THE NUMBER, RANGE, AND PROJECTION ANGLES OF THE DETECTION BEAMS VARY DEPENDING UPON DESIGN.

TYPICAL PIR COVERAGE PATTERN (CEILING MOUNTED)

DISC FLOOR BEAM PATTERN

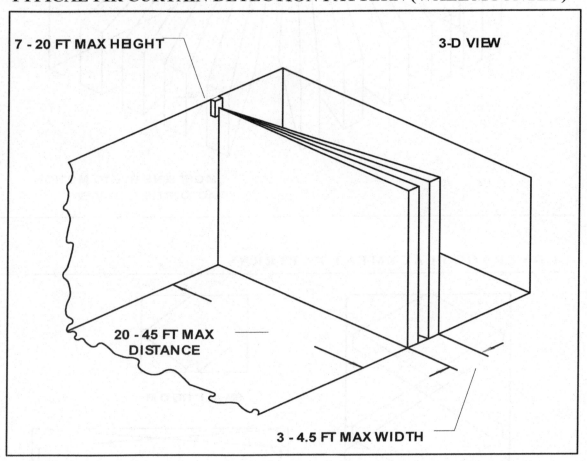

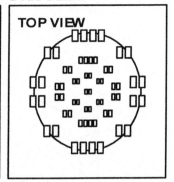

TYPICAL PIR CURTAIN DETECTION PATTERN (WALL MOUNTED)

PASSIVE INFRARED

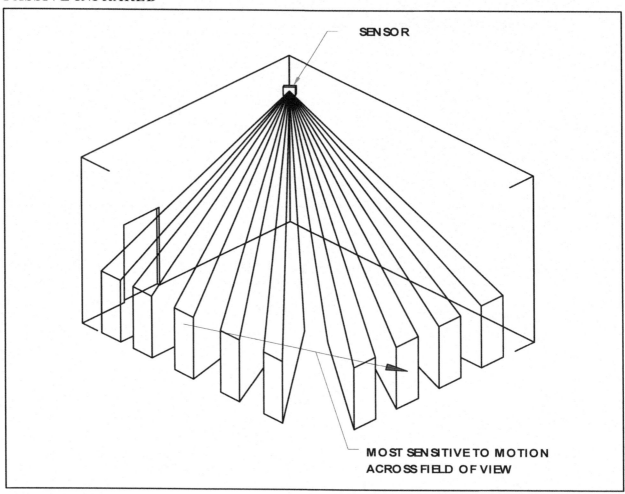

PIR COVERAGE/PLACEMENT PATTERNS

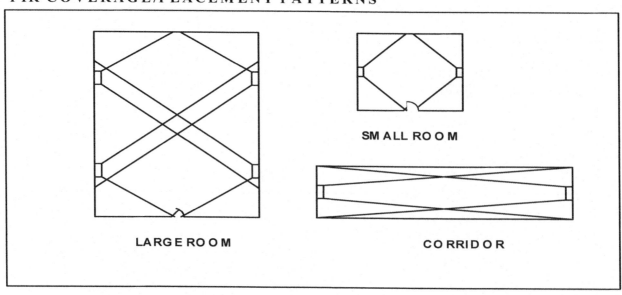

LARGE ROOM

SMALL ROOM

CORRIDOR

TECHNOLOGY REVIEW # 13

INTERIOR ACTIVE INFRARED

1.   Introduction: Interior active infrared sensors generate a curtain pattern of modulated infrared energy and react to a change in the modulation of the frequency or an interruption in the received energy. Both of these occurrences happen when an intruder passes through the protection zone.

2.   Operating Principle: Interior active infrared sensors are made up of a transmitter and receiver encased within a single housing unit. The transmitter uses a laser to create a detection zone. The laser plane is projected onto a special retro-reflective tape that defines the end/edge of the protection zone. Energy is reflected off the tape back to the receiver, which is located in the same housing unit as the transmitter. Upon reaching the receiver the energy passes through a collecting lens that focuses the energy onto a collecting cell, which converts the infrared energy to an electrical signal. The receiver monitors the electrical signal and generates an alarm when the signal drops below a preset threshold for a specific period of time. An intruder passing through the field of detection will interrupt the signal and temporarily cause the signal to fall below the threshold value.

3.   Applications and Considerations:

a.   Applications: Depending upon which type of tape is used as the reflective medium, coverage patterns can be between 15-25 feet wide by 17-30 feet long. In addition, the laser plane angle can be adjusted from 37° to 180°. This system has a high probability of detecting intruders. Speed or direction of the intruder, and the temperature of the environment, have no effect on detection characteristics.

b.   Conditions for Unreliable Detection: Dust or other particles collecting on the surface of the reflective tape will hinder the detection capabilities. The reflective tape must have no gaps and be continuous to ensure reliable detection, and the angle from the sensor to the ends or corners of the tape must not exceed 45°.

c.   Causes for Nuisance Alarms: The activation of an incandescent light which shines directly into the sensor itself will generate an alarm. Also, incandescent lights greater than 100 Watts (or sunlight) falling directly in line with the tape will be reflected back to the receiver with a magnitude significant for alarm signaling.

4.   Typical Defeat Measures: Avoidance of the projected laser plane. A knowledgeable intruder can deduce the field of the potential detection pattern from the location of reflective tape, and plan his movements to avoid detection.

ACTIVE INFRARED MOTION SENSOR

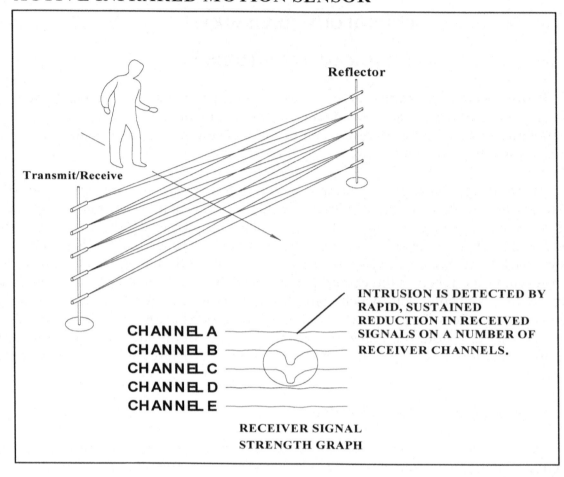

Reflector

Transmit/Receive

**INTRUSION IS DETECTED BY
RAPID, SUSTAINED
REDUCTION IN RECEIVED
SIGNALS ON A NUMBER OF
RECEIVER CHANNELS.**

CHANNEL A
CHANNEL B
CHANNEL C
CHANNEL D
CHANNEL E

**RECEIVER SIGNAL
STRENGTH GRAPH**

TECHNOLOGY REVIEW # 14

EXTERIOR ACTIVE INFRARED

1. **Introduction:** Active infrared sensors generate a multiple beam pattern of modulated infrared energy and react to a change in the modulation of the frequency, or an interruption in the received energy. Both of these occurrences happen when an intruder passes through the area covered by the beams.

2. **Operating Principle.** An active infrared sensor system is made up of two basic units, a transmitter and a receiver. One of the units is located at one end of the protection zone and the other at the opposite end of the zone. The transmitter generates a multiple frequency *straight line beam* to the remote receiving unit, creating an infrared "fence" between the transmitter and the receiver. Energy reaching the receiver passes through a collecting lens that focuses the energy into a collecting cell, which converts the infrared energy to an electrical signal. The receiving unit monitors the electrical signal and generates an alarm when the signal drops below a preset threshold for a specific period of time. An intruder passing through the field of detection will interrupt the signal and temporarily cause the signal to fall below the threshold value.

3. **Applications and Considerations:**

    a. **Applications:** Exterior active infrared sensors are line of sight devices that require the area between the two units to be uniformly level and clear of all obstacles/obstructions that could interfere with the IR signal. Low spots in the terrain will create "holes" in the surveillance pattern while obstacles/obstructions will disrupt the "coverage" pattern. Typically, active infrared sensors are used in conjunction with a single or double fence barrier which defines the perimeter to be covered. A sensor zone length can extend up to 1,000 feet.

    Precise alignment of the transmitter to the receiver is critical for reliable detection. The detection beam is relatively narrow and requires regular calibration/realignment. Detector misalignment could be caused by movements in the ground (e.g., earth tremors), objects hitting the unit (e.g., falling rocks, vehicles, falling trees) or even freezing and thawing of the ground.

    In areas where freezing ground or extreme winds are expected, the transmitter (Tx) and receiver (Rx) foundations should be installed deep enough to restrict movement/misalignment of the two units. In areas where the units are susceptible to being hit or jarred, protective barriers should be installed around them. Snow and grass around the Tx and Rx should be removed by hand or blower to prevent damage or misalignment of the units.

b. **Conditions for Unreliable Detection:** Weather conditions such as fog, heavy rain or severe sand/dust will attenuate the infrared energy and can affect the reliable detecting range. In areas where conditions like these are routine, another type of device should be considered, or the detection zone should be decreased to compensate for energy reduction.

c. **Causes for Nuisance Alarms:** Major causes of nuisance alarms are those that involve animal interaction with the protected area. Vegetation also can pose a problem if allowed to grow to a size its movement (caused by windy conditions) will generate an alarm.

4. **Typical Defeat Measures:** Since active infrared detectors are line of sight devices, the most common method of defeat is bridging/tunneling under the detection beams. For this reason it is recommended that any dips or gullies between the transmitter and receiver units/columns be filled in to make the area uniformly level. Another typical defeat measure is to use the Tx and Rx columns for support to vault over the detection beams. This can be prevented by overlapping the beam detection zones.

ACTIVE INFRARED MOTION SENSOR

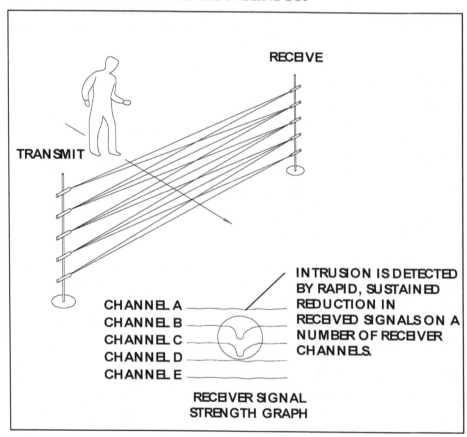

TECHNOLOGY REVIEW # 15

DUAL-TECHNOLOGY PASSIVE INFRARED / MICROWAVE

1.   Introduction: Dual-Technology Passive Infrared/Microwave sensors use a combination of both microwave and passive infrared technology in combination with *AND* logic to provide a lower False Alarm Rate (FAR) sensor than either of the sensors independently. This category of sensors are typically referred to as Dual-Tech.

2.   Operating Principle. In this type Dual-Technology sensor, a passive sensor (PIR) and an active sensor (Microwave) are combined into one unit. Both sensing elements are located in a single casing, and are connected electronically by using the *AND* Logic function. The areas of coverage for each sensor are similar in shape so the detection zone is uniform. Since the two sensors will not "sense" an intrusion detection precisely at the same instant, the system is designed to generate an alarm when both sensors produce an output in a pre-selected time interval. NOTE: The technical parameters and operating characteristics of each sensor are described in previous reviews (Tabs 6 and 12).

3.   Applications and Considerations:

    a.   Applications: The sensors can be installed along a perimeter line, a fence or a delineated buffer zone, or as a defense against intruders approaching a door or wall. To further enhance the probability of detection, image/video motion detection equipment can also be installed to survey the intrusion/approach zone. In addition to increasing the detection potential, this capability permits security personnel to assess the nature of the "intrusion/alarm" immediately and remotely.

    Although a dual-technology sensor does reduce the false alarm rate (FAR), it also reduces the probability of detection, since both sensors must have a positive detection before initiating an alarm. The mathematical probability of detection for the dual-tech unit is the product of the probability of detection for both individual units. For example, given a theoretical individual detection of rate 99 percent and 98 percent, the detection percentage for the Dual-Technology (*And logic* configuration) drops to 97.02 percent.

    b.   Conditions for Unreliable Detection: Since passive sensors have the greatest probability of detection when the intruder is moving transversely, and active sensors have the greatest probability of detection when the intruder is moving radially, the position of the sensor will dictate a positional trade-off that diminishes the sensor's detecting ability. Any condition that causes unreliable detection for the microwave sensor, as described in Tab 6 or the PIR sensor in Tab 12, can cause problems for the dual-tech sensor because the *AND* Gate Logic function needs signals from both sensors to generate an alarm.

Likewise, any environmental conditions that affect the performance of either sensor will reduce the effectiveness of the dual-tech. However, dual-technology sensors can be both cost effective (cheaper than purchasing two individual sensors) and FAR beneficial if employed in a predictable and/or controlled environment.

c.   **Causes for Nuisance Alarms:**   Nuisance Alarm Rate for the dual technology sensor is very low, however, a combination of environmental conditions (e.g. fluorescent lights, heater exhaust) may cause false detection. Environmental conditions that affect each sensor individually should be considered (compensated for) to keep from reducing effectiveness of the dual technology unit.

4.   **Typical Defeat Measures.** Knowledge of the dead spots in the detection pattern will permit an intruder to bypass all active regions. Short of this knowledge, extreme slow motion movement is difficult for microwave sensors to detect, and blocking or masking the infrared sensor's field of view can further decrease its sensitivity and reduce the probability of sufficient "heat" being detected by/focused on the PIR portion of the sensor. In addition, walking into the PIR sensor, rather than across its field of view, can reduce the detection capability of the sensor by not "breaking" the boundaries of the PIR detection beams.

TECHNOLOGY REVIEW # 16

FENCE VIBRATION

1.   **Introduction:** Fence vibration sensors mounted on fence fabric detect frequency disturbances associated with sawing, cutting, climbing or lifting of the fence fabric.

2.   **Operating Principle:** All of these type actions generate mechanical vibrations and/or stress in the fence fabric that are different from the vibrations associated with normal or natural occurring environmental activity, and typically have higher frequencies and larger amplitudes. Fence vibration sensors detect these vibrations by using either electro-mechanical or piezoelectric transducers. Signals from the transducers are sent to the signal processor to be analyzed. Upon arriving at the processor, frequencies uncharacteristic of intrusion are filtered out. Frequencies characteristic of intrusion are passed through the screening filter, thus triggering an alarm.

3.   **Sensor Types/Configurations:** There are two basic types of fence vibration sensors: Electro-mechanical sensors, whose signal processor has a pulse accumulation circuit that recognizes momentary contact openings of electromechanical switches; and Piezoelectric, whose signal processor responds to the amplitude, duration, and frequency of the transmitted signal.

    a.   **Electro-Mechanical Sensors:** Electro-mechanical sensors use either *mechanical inertia switches* or *mercury switches* to detect fence vibration or stress.

Mechanical-inertia switches consist of a *vibration sensitive mass* that rests on two or three electric contacts thus creating a closed circuit. The mass is movable and reacts to minute changes in the vibrations (frequencies) generated in the fence during a penetration attempt. The vibration disturbs the mass and is moved/separated from one or more of the contact points momentarily opening the circuit and creating an alarm. In some sensors the mass is intentionally constrained or restricted by some internal guides to ensure that only a significant vibration will cause movement, break the circuit and activate the alarm.

Mercury switches consist of a glass vial containing a small amount of mercury with a set of normally "open" electrical contacts located in close proximity, but not touching or immersed in the mercury. An impact-disturbance of the fence fabric causes the mercury to be displaced from its normal resting position, making momentary contact with one of the electrical contacts and creating an alarm.

b. **Piezoelectric Sensors:** Piezoelectric sensors convert the mechanical impact forces generated during an intrusion attempt into electrical signals. Unlike the open/close signal generated by electro-mechanical sensors, piezoelectric sensors generate an analog signal that varies proportionally in amplitude and frequency to the vibration activity on the fence fabric. These signals are sent to the signal processor for evaluation, where they first pass through a filter that screens out signals uncharacteristic of intrusions. The signal processor then interprets the remaining signals to determine if sufficient activity has occurred to warrant an alarm.

4. Applications and Considerations:

a. **Applications:** Fence vibration sensors perform best when mounted directly to the fence fabric. Each sensor is connected in series along the fence with a common cable to form a single zone of protection. The sensor zone lengths have a recommended range of 300 feet.

Vibration sensors are the most economical fence sensor and the easiest to install. The sensors have a high probability of detecting intrusion and work well protecting properly installed and maintained fence lines.

In-ground vibration (seismic) sensors installed adjacent to the perimeter fence (in a controlled zone within the overall protected area) can provide additional detection capability protection in case the vibration sensors mounted on the fence are bypassed by tunneling or careful climbing.

Another type of enhancement focuses on adding information about the prevailing weather conditions to increase or decrease the sensitivity of the processor. A weather sensor station can be mounted on the fence line to feed information to a field processor. The field processor then adjusts vibration alarm sensitivity based on inputs from the weather station to ensure an effective sensitivity range is maintained.

Mounting volumetric motion detection devices (microwave, active infrared) along the perimeter of the fence will also enhance detection reliability. Determining which volumetric (area monitoring) device to use will depend greatly on the environment, terrain and length of the fence line.

Because vibration sensors are prone to activation from all types of vibrations, additional sensing equipment is frequently added to the processor capability to reduce false activations. One type of enhancement is the pulse count accumulator circuit. With this device, sensitivity is determined by a number of "pulses" required to create an alarm. A pulse is a specific amplitude of activity occurring due to fence stress or vibration associated with cutting chain links or climbing the fence fabric. A minimum number of pulses is required during a preset period of time before an alarm is generated.

b. **Conditions for Unreliable Detection:** Proper installation and spacing of sensors is critical to reliable detection. Poor quality fences with loose fabric can create too much background activity (flexing, sagging, swaying), initially generating false alarms and eventually transmitting little reliable intrusion activity. Likewise, adverse weather conditions can cause sensitivity settings above/below what is required for reliable detection to occur. Fence corners pose particular challenges for readily detecting intrusion vibrations, because of the increased bracing of the fence posts and more solid foundations typically used at a corner or turn-point.

c. **Causes for Nuisance Alarms:** Shrubbery and tree branches as well as animals and severe weather that come in contact with the fence can cause the fence to vibrate, triggering the sensors to react. *In areas with high wind or numerous animal interactions with the fence line, vibration sensors should not be used.* Vibration sensors should only be used in areas/circumstances where natural or man-made environmental vibrations are minimal or non-existent. Vibration sensors are not satisfactory nor are they reliable in areas/situations where high vibrations are likely to be encountered, such as in close proximity to construction sites, railroad tracks/yards or highway and roadway activity.

5. **Typical Defeat Measures:** The most common defeat method is to avoid contact with the fence by bridging it. Overhanging trees and structures can assist the intruder in this regard. Similarly, cars, buses, trucks, equipment or storage containers positioned/parked next to the fence can serve as platforms for jumping/bridging the fence. Although less common, deep tunneling, if accomplished without contacting the fence supports, will allow an intruder to bypass a fence mounted vibration sensor system.

VIBRATION FENCE SENSOR

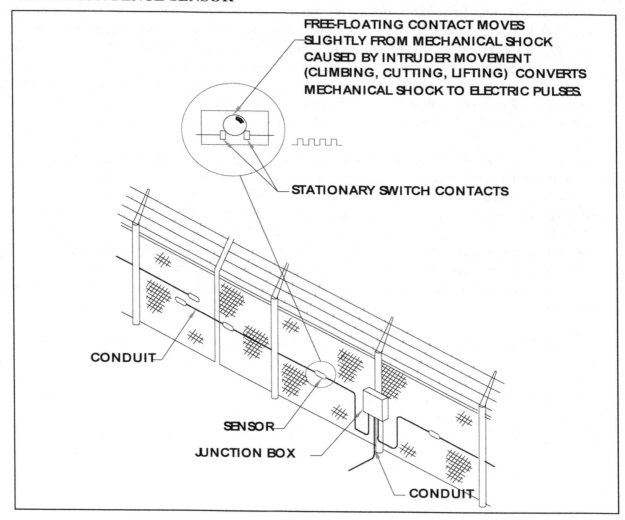

FREE-FLOATING CONTACT MOVES SLIGHTLY FROM MECHANICAL SHOCK CAUSED BY INTRUDER MOVEMENT (CLIMBING, CUTTING, LIFTING) CONVERTS MECHANICAL SHOCK TO ELECTRIC PULSES.

STATIONARY SWITCH CONTACTS

CONDUIT

SENSOR

JUNCTION BOX

CONDUIT

TECHNOLOGY REVIEW # 17

ELECTRIC FIELD

1.   Introduction: Electric field sensors generate an electrostatic field between/around an array of wire conductors and an electrical ground. Sensors in the system detect changes or distortion in the field. This can be caused by anyone approaching or touching the fence.

2.   Principle of Operation: The E-field sensor consists of an alternating current field generator which excites a field wire (two or more sensing wires), around which an electro-static field is created and an amplifier signal processor which detects changes in the signal amplitude of the sensing wires. The alternating current on the field wire creates an electrostatic field in the air between the field wire to ground. When an intruder enters the "field", large amounts of the electric charge flow through the intruder due to the human body disrupting the field. The processor detects this change and generates an alarm.

To reduce false alarms, the signal goes through a filter which rejects high frequencies caused by wind vibration and low frequencies caused by objects striking the fence wires. However, the filter allows frequencies associated with intrusion characteristics to continue to the processor. At the processor, three conditions must be met to signal an alarm: the signal amplitude must exceed a preset value that discriminates small animals, the frequency must be in a range that is associated with humans, and the signal must persist for a set period of time. Once these conditions are met, the processor signals an alarm.

3.   Applications and Considerations:

   a.   Applications: E-field wire configurations are mounted on free-standing posts or chain link fences. All the wires are mounted parallel to each other and to the ground, thereby achieving uniform sensitivity along the fence length. Springs are used at the connectors to ensure tension reducing vibrations caused by wind.

An advantage that an E-field sensor has over other fence sensors is the self adjusting circuit, located in the processor, that rejects wind and ambient noise. This circuit not only requires the amplitude of an intrusion attempt to exceed a preset level, but also for a preset period of time. The E-field sensor has an extremely low Nuisance Alarm Rate. In some cases bridging and tunneling can be detected, depending on how close the disturbance activity is to the sensor. Sensor zone length can extend up to 1500 feet. The

E-field sensor should be considered if bridging or tunneling are expected intrusion tactics. Other fence sensors (vibration, taut wire) can be added to provide a higher level of detection probability.

b. **Conditions for Unreliable Detection:** Adverse weather conditions such as rain and snow can create problems, as can lightening storms. In addition, vegetation and animal movement along the fence line can cause the sensors to react. Large spacing between wires should be avoided, as it is possible to move between the wires without causing an alarm if sufficient space exists.

NOTE: Although Electronic Magnetic Interference (EMI) is not normally a major factor, interference difficulties can arise in situations where multiple systems are deployed in a congested area, unless different frequencies are used by each sensor.

c. **Causes for Nuisance Alarms:** Anything causing excessive fence vibration such as weather, birds, and animals will contribute to nuisance alarms. Overgrown vegetation coming in contact with the fence line can also be a problem and should be avoided by keeping grass and shrubbery clear of the fence.

4. **Typical Defeat Measures:** Although electric field sensors provide some means of detecting underground intrusion activity because of disturbances in the electric field, the sensor field can be bypassed by deep tunneling (6 feet or more) or bridging over the fence.

ELECTRIC FIELD DETECTION CONFIGURATION/PATTERNS

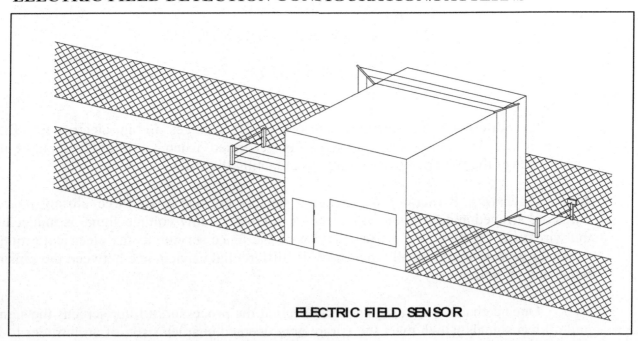

ELECTRIC FIELD SENSOR

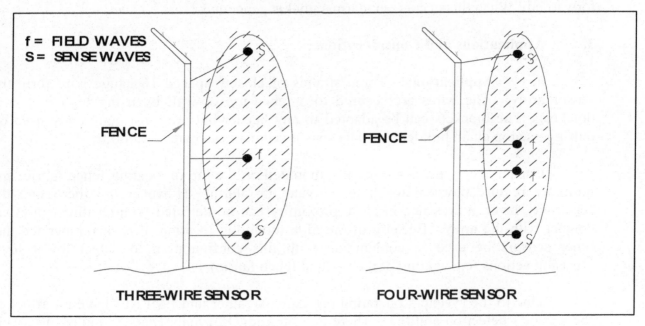

f = FIELD WAVES
S = SENSE WAVES

FENCE

THREE-WIRE SENSOR

FENCE

FOUR-WIRE SENSOR

TECHNOLOGY REVIEW # 18

CAPACITANCE

1.   Introduction: Capacitance sensors detect changes in an electrostatic field created by an array of wires. A signal is generated when an intruder changes the capacitance of the field by approaching or contacting the wires.

2.   Operating Principle: Capacitance sensors consisting of three closely spaced wires are arrayed and installed on the top of a fence. A low voltage signal is induced in the wire array creating an electrical field with the fence serving as the electrical ground. A sensor processor continually measures the differential capacitance between the sensing wires and ground.

Once a change in the signal is detected at the processor, a filter screens the signal and allows signals which meet the parameters deemed characteristic of an intruder to be forwarded. When this occurs, an alarm signal is generated.

3.   Applications and Considerations:

a.   Applications: Three strands of closely spaced 16 gauge wire form the sensor array. The wires are secured to a fence top or wall by using high dielectric brackets. The brackets can be adapted to any barrier but are most commonly used on outriggers atop chain link fences. The sensor segment can extend 1,000 ft.

Capacitance sensors are usually mounted on the top of existing fence fabric, and normally require physical touch to activate the alarm. However, by increasing the sensitivity level, a presence in close proximity can be detected without direct physical contact with the array. Because of the high mounting location, it is recommended that other sensors be used in conjunction with this configuration to detect lower level intrusion actions (e.g. cutting of the vertical fence fabric).

Due to the system's operating principle, weather and EMI/RFI have no affect on the sensor's detection ability.   There is a high level of maintenance required to assure the capacitance characteristics of the fence are always adjusted.

b.   Conditions for Unreliable Detection: Unreliable detection may occur from vibrations caused by weather and animals, interpreted as intrusion attempts. Vegetation coming in contact with the fence will change the capacitance, thereby affecting the detection characteristics. To avoid this, proper landscaping maintenance of the fence line must be done (grass cut, trees removed, shrubs removed).

    c.   **Causes for Nuisance Alarms:**   Animals, such as birds and squirrels, contacting the fence will generate an alarm. This can be reduced by removing possible food sources (shrubs, grass).

    In addition, blowing debris, or anything making physical contact that changes the characteristics of the fence, may generate an alarm condition. Inducing tension with springs at the termination points can reduce this possibility.

4.   **Typical Defeat Measures:** Bypassing the sensors by tunneling or bridging is a method of defeat.

CAPACITANCE SENSOR APPLICATIONS

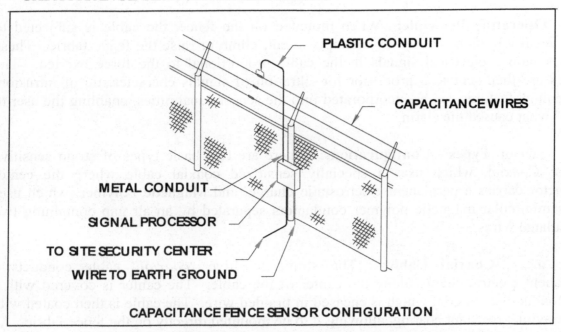

PLASTIC CONDUIT

CAPACITANCE WIRES

METAL CONDUIT

SIGNAL PROCESSOR

TO SITE SECURITY CENTER

WIRE TO EARTH GROUND

CAPACITANCE FENCE SENSOR CONFIGURATION

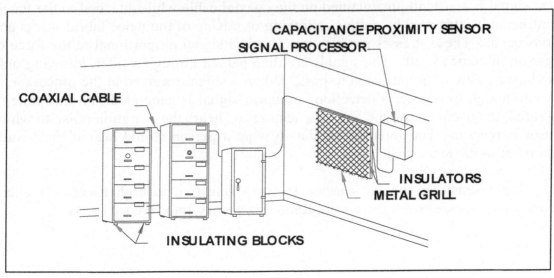

CAPACITANCE PROXIMITY SENSOR
SIGNAL PROCESSOR

COAXIAL CABLE

INSULATORS

METAL GRILL

INSULATING BLOCKS

TECHNOLOGY REVIEW # 19

STRAIN SENSITIVE CABLE

1.   Introduction: Strain sensitive cables are line sensors that use electric energy as a transmission and detection medium. The line sensors maintain uniform sensitivity over the entire length of the protection zone. The cable runs from the signal processor to an end-of-line resistor, which guards against cutting, shorting or removal of the cable from the processor.

2.   Operating Principle: When mounted on the fence, the cable is subjected to mechanical vibrations caused by attempts to cut, climb or raise the fence fabric. These stresses induce electrical signals in the cable proportional to the force exerted. The signals are then sent to a processor for filtration of signals characteristic of intrusion. "Listening" features can be incorporated into the sensor capabilities, enabling the user to "hear" what caused the alarm.

3.   Sensor Types / Configurations: There are two basic types of strain sensitive cables: Coaxial, which uses a specially sensitized coaxial cable where the center conductor carries a permanent electrostatic charge, and Magnetic Polymer, which uses two semicircular magnetic polymer conductors separated by an air gap containing two uninsulated wires.

a.   Coaxial Cable: The strain sensitive *"coaxial"* cable conducts a permanent electric charge along the center of the cable. The center is covered with a non-conductive material which is encased in braided wire. The cable is then coated with an ultraviolet resistant coating, allowing it to be mounted directly on the fence fabric. An electrical signal is constantly maintained on the coaxial cable while attached to the fence. When intrusion is attempted by cutting, climbing or raising of the fence fabric, stress and vibrations occur. These stresses produce an electrical signal proportional to the force of the stress on the fence itself. The signals are then passed through a filter, allowing only signals characteristic of an intrusion to pass. When a signal received at the processor, is significant enough to register a detection, an alarm signal is generated. At that time, if incorporated, the user can use the listening feature to "hear" the vibration noise to which the sensor is reacting. The sound is similar to what a person would hear if their could press their ear to the fence post.

NOTE: Strain sensitive coaxial sensors are very sensitive to high Electro-Magnetic Interference (e.g., power substations) and Radio Frequency Interference (RFI).

b. **Magnetic Polymer:** Strain sensitive *"magnetic polymer"* cable sensors function as poles of a linear magnet. This is done by pairing two semicircular, magnetic, polymer conductors and separating them by an air gap. Two insulated wires run between the polymers. Parallel to these wires are two uninsulated wires free to move in the air gap between the magnetic field created by the polymer conductors. Vibration and stress on the fence fabric cause the active conductors (uninsulated wire) to move within the air gap. When this movement takes place, slight electric signals are generated and sent to the signal processor. The processor compares the signal and generates an alarm if it's outside the pre-calibrated parameters. Processors are available that "learn" from normal fence fluctuations and revise the data, thus enhancing the performance of the system.

The *"magnetic polymers"* multiple conductors form a symmetrical and balanced pair configuration which makes the cable essentially unsucceptible to both Electro-Magnetic Interference and Radio Frequency Interference. Its low impedance creates higher signal-to-noise ratios, which provide more finite signal processing. The magnetic polymer cable also functions as a transducer microphone, and can have a "listening" operation implemented in the system, allowing the user to audibly interpret the activity taking place at the fence line.

4. **Applications and Considerations:**

a. **Applications:** The cable is shielded with an ultraviolet resistant coating, and is designed to be mounted directly to the fence fabric. Strain-sensitive cables should be installed using ties halfway between the bottom and the top of the fence. Also, stainless steel wire ties (vs. plastic ties) should be used to prevent silent removal by burning (e.g. cigarette butane lighter). Sensor zone lengths can extend up to 1,000 feet, however, a quality fence and stable installation are necessary for reliable detection.

Intrusion detection probabilities can be enhanced by mounting volumetric motion detection devices (microwave, active infrared) along the perimeter of the fence line. Other fence sensors (e.g., electric field) can be employed in tandem to provide increased detection possibilities.

In addition, Video Motion Detection cameras mounted to view the protected area, will provide another layer of detection potential while also allowing security personnel to assess the alarm visually.

b. **Conditions for Unreliable Detection:** Poor fence construction and/or unstable installation and lack of proper maintenance will decrease detection potential.

c. **Causes for Nuisance Alarms:** Severe weather can present nuisance alarms; however, with proper calibration and installation most normal weather problems can be avoided. Routine animal-caused alarms can be filtered out by employing the listening device to determine legitimate intrusion signals from false ones.

5. **Typical Defeat Measures:** As with other fence-based sensors, bridging over or tunneling under, the fence will bypass the detection system. Also, an intruder conscious of the system installation and configuration may be able to climb the fence without detection.

STRAIN - SENSITIVE CABLE (COAXIAL)

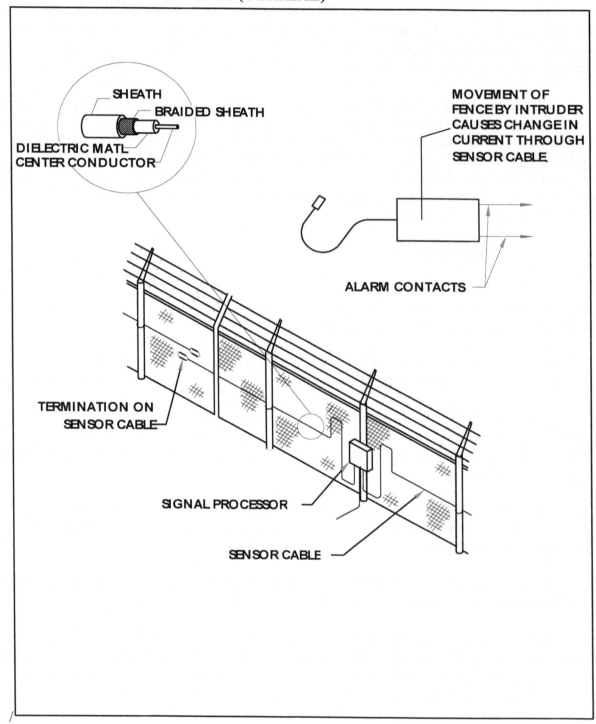

STRAIN SENSITIVE CABLE (MAGNETIC)

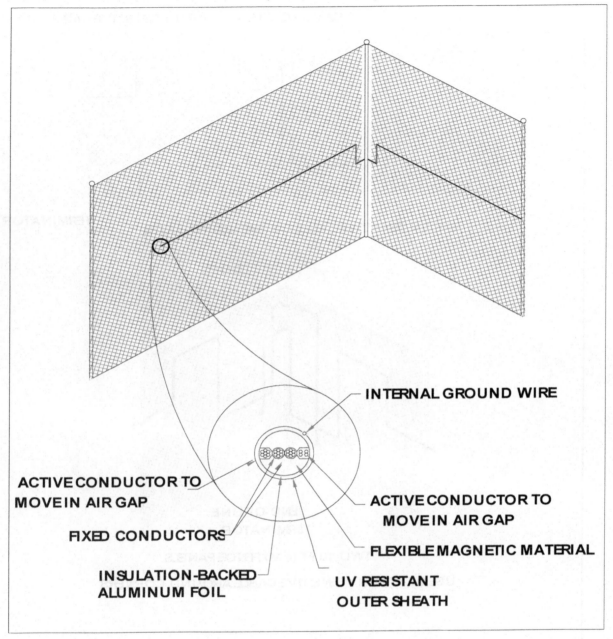

INTERNAL GROUND WIRE

ACTIVE CONDUCTOR TO
MOVE IN AIR GAP

ACTIVE CONDUCTOR TO
MOVE IN AIR GAP

FIXED CONDUCTORS

FLEXIBLE MAGNETIC MATERIAL

INSULATION-BACKED
ALUMINUM FOIL

UV RESISTANT
OUTER SHEATH

STRAIN - SENSITIVE CABLE (APPLICATIONS)

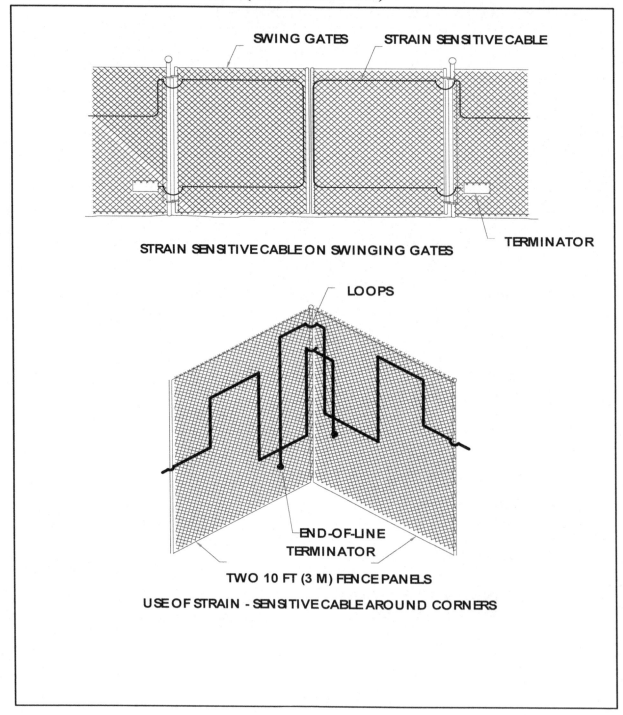

STRAIN SENSITIVE CABLE ON SWINGING GATES

TWO 10 FT (3 M) FENCE PANELS

USE OF STRAIN - SENSITIVE CABLE AROUND CORNERS

TECHNOLOGY REVIEW # 20

FIBER OPTIC FENCE

1.   Introduction: Fiber optic sensors use light rather than electricity for transmission and detection. Fiber optic cable is ideal for incorporation into existing fences, or it can be used as stand alone fencing.

2.   Operating Principle: Optical fiber is a fine, strong strand of glass or other optical medium. It is often called a wave guide because the optical fiber guides light waves from a light source at one end to a detector at the other end of the fiber. In operation, light is pulsed through the fiber in a manner similar to an electric signal through a wire. Fiber optics, however, offer several distinct advantages over other conductive materials. Optical fiber is immune to electrical interference and Electrical Magnetic Interference (EMI) disruption. It is intrinsically safe and uses very stable equipment, making it highly reliable overall. The power of light passing though a fiber optic is measured in decibels (dB) of light energy, with the fiber optic absorbing approximately 3 dB of light per kilometer, allowing use of the system over great distances.

3.   Sensor Types / Configurations: Depending on the processor used, two basic types of fiber optic sensors can be employed: fiber optic *continuity,* which requires the fiber optic strand to be broken to initiate an intrusion alarm, and fiber optic *microbending*, which detects alterations in the light pattern caused by movement of the fiber optical cable.

     a.   Fiber Optic Continuity: The fiber optic continuity sensor is similar to any closed loop device. As long as the sensor cable remains intact, with light passing from the transmitter to the receiver, no alarm is transmitted. If the cable is broken, the signal transmission ceases and an alarm is generated.

     In one form a composite strand sensor is combined with a fiber optic continuity sensor within a barbed steel tape. An installed system appears similar to a taut wire installation, however, it does not require the mechanical activation of switches, making preventive maintenance and repair of the system more affordable. The fiber optic barbed tape can be used as a free standing fence or it can be applied to an existing fence. It can also be attached to walls and buildings.

     b.   Fiber Optic Microbending / Disturbance: As the name implies, the fiber optic cable must be bent or disturbed in some way, to affect the wave guide of the light being transmitted and thereby signaling a disturbance. Detection is a function of stress on the fence fabric.

The fiber optic cable acts as a line sensor when installed on the fence fabric itself. The system contains an electro-optics unit, which transmits light using an LED for the light source. The light travels through the fiber optic and is picked up by the detector, which is very sensitive to slight alterations in the transmission caused by vibration or strain on the fence. When an adequate alteration in the light pattern takes place, an alarm signal is generated.

4.   Applications and Considerations:

      a.   Applications: Fiber optic fence sensors should be mounted directly on, or woven into, the fence fabric. A quality and stable installation of the fence is necessary for reliable detection. Freedom from rattles, clanks, knocking sounds, and vibration/stress activity maximizes line sensor quality. The more activity that exists around the fence, the lower the sensitivity setting for the system, and the less likely the system will detect an intruder.

      To enhance the potential for intrusion detection, *In-ground sensors* can be installed within the protected fence area providing another level of detection. Video motion detection cameras mounted outside or inside the protected fence area can increase the intrusion detection potential, and allow security personnel to assess the intrusion zone visually. An additional way to enhance the security of a fiber optic fence is to mount a volumetric motion detection device (e.g. microwave, active infrared) along the perimeter of the fence.

      b.   Conditions for Unreliable Detection: Poor fence quality (stability) is the most common condition for unreliable detection. Loose fence fabric and poor stability cause the sensitivity setting for the fence to be calibrated lower than preferred. This makes the system less likely to detect an intruder. When properly installed on a good quality, stable fence or installed in a taut wire-like configuration, the system is very stable.

      c.   Causes for Nuisance Alarms: Although the system is impervious to transient voltage/lightning strikes, system problems can be created by Radio Frequency Interference (RFI), Electro-Magnetic Interference (EMI), extreme changes in temperatures and blowing debris. Although most normal weather conditions can be screened out by the alarm processor, extreme weather turbulence that disturbs or damages the optical fiber cable can create nuisance alarms. In addition, animal activity coming in contact with the fence can be interpreted as human activity, falsely signaling an intrusion attack.

5.   Typical Defeat Measures: Bridging or tunneling will bypass the fence and, therefore, bypass the sensor. Careful or assisted climbing, particularly at the more rigid turn points, may not produce the activity level required for alarm activation.

FIBER OPTIC CABLE

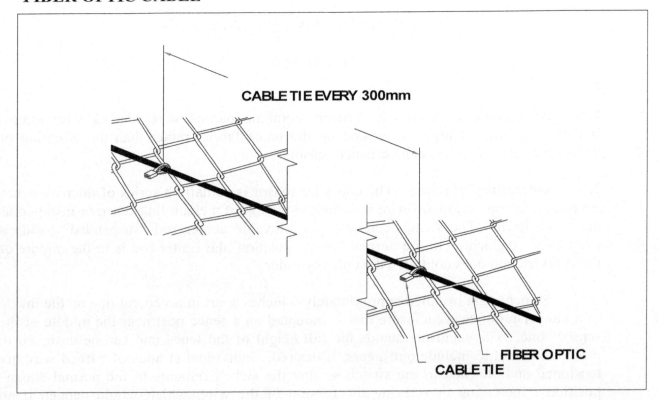

CABLE TIE EVERY 300mm

FIBER OPTIC
CABLE TIE

TECHNOLOGY REVIEW # 21

TAUT WIRE

1.   Introduction: Taut wire sensors combine barbed wire fencing with micro-switches to detect changes in tension on the fence fabric, rather than the vibration or stress associated with fence disturbance sensors.

2.   Operating Principle: The taut wire sensor is actually a series of microswitches connected to tensioned barb wire installed on the top of a chain link fence or installed as the fence itself.   The switch consists of a movable center rod "suspended" inside a cylindrical conductor.   In the normal "open" position, the center rod is in the middle of the cylinder, and does not touch the outer cylinder.

Switches are installed approximately 6 inches apart in a vertical line on the inside of a tamper-proof case/enclosure that is mounted on a fence post near the middle of the sensor zone.   The enclosure stands the full height of the fence and can be designed to project outward to include outriggers, if desired.   Individual strands of barbed wire are tensioned and attached to the switch so that the switch remains in the normal "open" position.   Increasing or relaxing the tension of the wire, which would happen if an intruder attempted to climb, spread or cut the wire, causes the inner center rod to come in contact with the outer cylinder, "closing" the contact and initiating the alarm sequence.

A unique, critical feature of the switch is a pliable, plastic support for the switch housing.   This material exhibits cold-flow properties that allow the switch to always assume a neutral position when acted upon by gradual external forces such as fence settling or freezing/thawing temperatures.   This feature prevents the switch from becoming pre-stressed thereby altering the intended sensitivity of the sensor.

The sensor is not overly susceptible to wind conditions, and a firm pull/force is needed to activate a switch. The Taut Wire design is intended to activate an alarm on the first switch contact, as this may be all that is indicative of an intrusion attempt or penetration action. Regular tensioning (maintenance) of the system is critical to ensure the system performs as intended.

3.   Sensor Types/Configurations: Taut wire sensors can be mounted in two different configurations: (a) on top of an existing fence in conjunction with barbed wire outriggers to provide protection from climbing, or (b) as the fence fabric itself.

a.   Outriggers: In situations where the taut wire sensors are mounted on the top of an existing fence using barbed outriggers, they are targeted to deter/detect intrusion attempts by climbing. In this configuration, the sensors will have little affect on the cutting of the lower fence fabric, potentially allowing undetected access. Because of this vulnerability, it is recommended that another type of sensor (e.g. vibration), which can be mounted on the fence fabric, be used in conjunction with the taut wire system to detect cutting or raising of the fence fabric in the lower section of the fence.

b. **Fence Fabric:** If mounted as the fence fabric, strands of barbed wire in a single zone are supported at each fence post, except the switch assembly post, by a supporting bar. The supporting bars loosely support the strands of barbed wire, allowing them to move freely to activate the taut wire switches.

NOTE: The combination of these two techniques provides an integrated barrier that detects cutting, climbing, and raising of the fence fabric. An advantage to this method is its high reliability, low False Alarm Rate (FAR)and low Nuisance Alarm Rate (NAR).

4. **Applications and Considerations:**

a. **Applications:** Taut wire sensors are used to protect perimeter fence lines. They are one of the most expensive fence sensor systems, because of the laborious installation and maintenance time required. Taut wire sensors are very reliable, and provide a high probability of detection and an extremely low false alarm rate. Because of these features, taut wire sensors are usually installed at high risk facilities. However, tedious, regular tensioning of the system is required to ensure the system performs as intended.

Exertion needed on the wire for activation is substantial, therefore, weather is not a factor in consideration for this sensor. Typically, small animals do not pose a threat for false alarms either, because of the magnitude of a 35 pound force needed for activation of the sensor.

To enhance the system, in-ground sensors can be installed inside the protected fence area, providing protection in the event the taut wire sensors are bypassed by tunneling or bridging. Furthermore, mounting volumetric motion detection devices (microwave, active infrared) along the perimeter of the fence will also increase the probability of detection. However, determining which volumetric device to use will depend greatly on the environment, terrain and length of the fence line. In addition, video motion detection cameras mounted outside or inside the protected fence area can provide a second level of security while also allowing personnel to assess the alarm quickly.

b. **Conditions for Unreliable Detection:** The system is one of the more reliable fence-based detectors, as it is less susceptible to environmental conditions and small animals. However, improper maintenance (tensioning) of the sensors can cause conditions for unreliable detection.

c. **Causes for Nuisance Alarms:** Medium to large animals that "push" the fence while grazing or nesting can generate an alarm.
NOTE: Because of the expense of taut wire systems, they are usually installed at high risk facilities where sterile areas prevent unintentional contact with the fence itself.

5. **Typical Defeat Measures:** Tunneling or bridging the fence itself. Tunneling is most likely to occur at a mid-point between fence posts in relatively soft ground. Bridging can occur anywhere along the fence line, with the most likely locations being those that are not under regular observation or provide the greater degree of concealment during the approach.

TAUT WIRE FENCE SENSOR

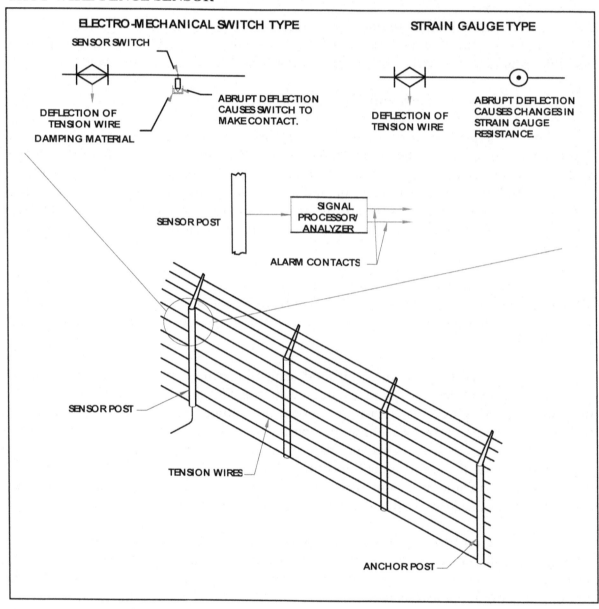

MICROWAVE AND TAUT WIRE SENSOR COMBINATION

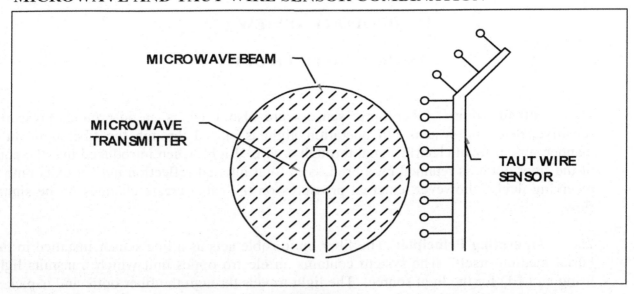

MICROWAVE BEAM

MICROWAVE
TRANSMITTER

TAUT WIRE
SENSOR

PORTED COAXIAL CABLE AND FENCE SENSOR COMBINATIONS

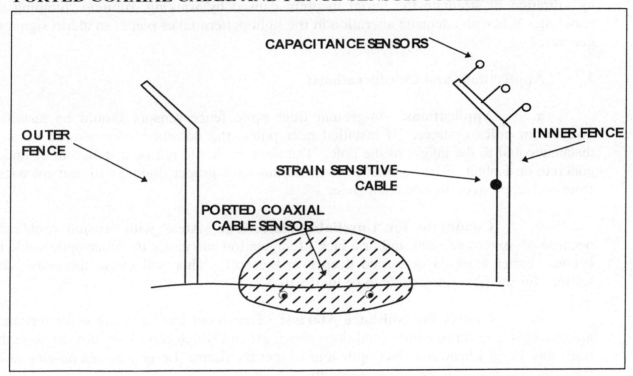

CAPACITANCE SENSORS

OUTER
FENCE

INNER FENCE

STRAIN SENSITIVE
CABLE

PORTED COAXIAL
CABLE SENSOR

TECHNOLOGY REVIEW # 22

IN-GROUND FIBER OPTIC

1.   Introduction: Fiber optic sensors are also used as an in-ground, pressure-sensitive, detection system.   In operation, light is pulsed through the fiber optic in a manner similar to an electric signal through a wire. Light, when introduced into the core of the fiber optic, is retained by a process of total internal reflection until it exits onto a receiving device, however, external pressures on the cable create changes in the signal flow.

2.   Operating Principle: The fiber optic cable acts as a line sensor installed in the burial medium itself. The system contains an electro optics unit which transmits light using an LED for the light source. The light travels through the fiber optic and is picked up by the detector, which is very sensitive to slight alterations in the transmission caused by vibration or strain in the burial medium caused by walking, running, jumping or crawling. When an adequate alteration in the light pattern takes place, an alarm signal is generated.

3.   Applications and Considerations:

    a.   Applications: In-ground fiber optic fence sensors should be installed away from poles or trees. If installed near poles, the detection zone should be at a distance equal to the height of the pole. The sensors should not be installed in or under concrete or asphalt. The installation area should have proper drainage to prevent water from collecting over the detection zone.

    b.   Conditions for Unreliable Detection: Areas with erosion problems, because of extensive rains and/or a lack of vegetation can cause the fiber optic cable to become either exposed or buried deeper in the soil. This will cause the sensitivity settings for the fiber optic cable to be ineffective.

    c.   Causes for Nuisance Alarms: Tree roots can be a cause for nuisance alarms. This is because windy conditions above ground which can cause movement in the roots and in turn bend the fiber optic and trigger an alarm. Large animals passing over the detection zone can also generate alarms.

4.   Typical Defeat Measures. Bridging over the sensors will bypass the system.

NOTE: It is recommended that a volumetric sensor (e.g., microwave) be used in conjunction with a fiber optic system to enhance detection probability.

IN-GROUND FIBER OPTIC SENSOR

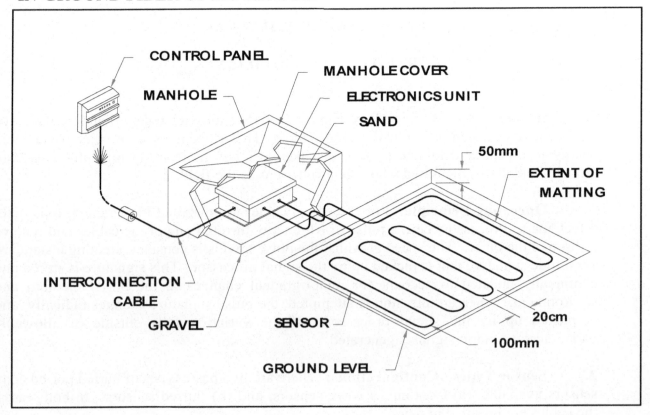

TECHNOLOGY REVIEW # 23

PORTED COAX BURIED LINE

1.   Introduction: Ported Coax Buried Line Sensors are coaxial cables that have small, closely spaced holes in the outer shield. These openings allow electromagnetic energy to escape and radiate a short distance. Emissions from these cables create an electric field that is disturbed when an intruder enters the field.

2.   Operating Principle: Ported coaxial cables are installed in pairs, approximately 5 feet apart. Processors emit a pulse of RF energy through one of the cables and receive it through the other. The speed at which the pulse travels is constant, creating a standard amplitude signature that is picked up by the signal processor. This signature is stored and continually updated to account for minor/gradual changes in the burial medium and environment. When an intrusion is attempted, the pulse signature changes radically, and is picked up by the signal processor. If the variation falls outside of allowable parameters, an alarm signal is generated.

3.   Sensor Types / Configurations: There are two basic types of buried ported coax sensors available: (a) Continuous wave sensors, and (b) Pulsed sensors. In both cases, the cables are installed in pairs.

    a.   Continuous Wave: With Continuous Wave sensors the RF energy is transmitted simultaneously by both cables and received by the opposite number. The energy emission is constant, thus creating a detection zone above ground with a continuous surface. When an intruder enters the detection zone, the electric field is disturbed, signaling the processor to generate an alarm.

    b.   Pulse: Pulse sensors emit a pulse of RF energy through one cable and receive it through the other. The speed at which the pulse travels is constant, creating a standard amplitude signature that is picked up by the signal processor. This signature is stored and continually updated to account for slow/small changes in the burial medium and environment. When an intrusion is attempted, the pulse signature changes and is picked up by the signal processor. If the variation falls outside of allowable parameters, an alarm signal is generated.

4.   Applications and Considerations:

    a.   Applications: The cables are buried approximately 9 inches below the surface of the ground, depending on the soil density, creating an electric field approximately 3-4 feet above the ground that extends 9-12 feet wide. The variation in zone size depends on cable separation and the characteristics of the burial medium. With this sensor cable, zone length can extend up to 500 feet.

Routing the cables underneath chain link fences should be avoided. If metallic pipes or cables must be routed through the sensor cable field, they should be buried at least 3 feet below the ported coaxial cable. When installing the cables along or near fence lines, the cables must be installed between 6 and 10 feet from the fence to avoid distortions and to reduce potential false alarms caused by the motion of the fence fabric disrupting the detection field. A video motion detection system can be used to complement the cable sensor and provide security personnel with the capability to assess alarm locations quickly and safely.

   b.   Conditions for Unreliable Detection: Because of the limited height of the detection zone, sites that experience heavy snowfall are prone to unreliable detection. Also, burial mediums that have drain ducts located beneath the buried cables will pose a problem if ducts are not constructed of metal. Wind disturbance of standing water over the cables also causes erroneous signals, therefore, the burial zone should be graded to provide immediate runoff and good drainage.

NOTE: Ported Coax sensors are affected by high EMI from sources such as large electrical equipment or electrical sub stations and should not be used in close proximity to these type installations.

   c.   Causes for Nuisance Alarms: Movement of nearby metallic fence fabric, vehicles and signs, as well as organic objects (e.g., people, medium to large animals, medium to large vegetation), can cause alarms. Individual small animals typically do not have the magnitude to effect the system, however, a congregation of small animals can generate an alarm.

5.   Typical Defeat Measures: Bypassing the area by bridging over the detection zone is the principle method employed.

BURIED PORTED CABLE SENSOR

BURIED PORTED CABLE SENSOR RESPONDS TO INTRUDERS MASS AND VELOCITY IN MOVEMENT THROUGH THE SENSOR FIELD.

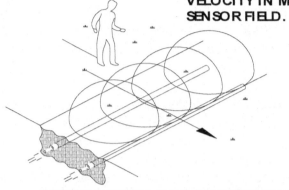

BURIED PORTED CABLE SENSOR SIZE AND SHAPE OF DETECTION PATTERN IS DEPENDANT UPON CABLE SEPARATION, BURIAL DEPTH, AND SOIL CONDUCTIVITY LEVELS.

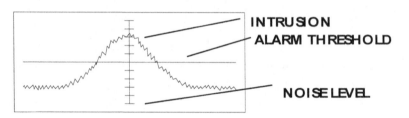

INTRUSION

ALARM THRESHOLD

NOISE LEVEL

BURIED PORTED CABLE SENSOR RESPONSE GRAPH

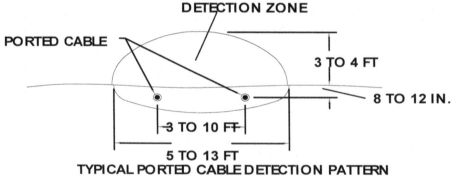

DETECTION ZONE

PORTED CABLE

3 TO 4 FT

8 TO 12 IN.

3 TO 10 FT

5 TO 13 FT

TYPICAL PORTED CABLE DETECTION PATTERN

TECHNOLOGY REVIEW # 24

BALANCED BURIED PRESSURE

1.   Introduction: A Balanced Buried Pressure line sensor is an in-ground system that detects vibrations and seismic energy. These energy waves are typically caused by personnel, animal or vehicular movement across the surface of the ground in which the sensors are installed.

2.   Operating Principle: Pressure line sensors consist of pressurized, closed end, pliable tubes or hose segments filled with water or an antifreeze-like solution. Usually two sensors tubes are used per zone. The zone size will vary depending on soil density and composition, and the nature of any surface material. The tubes are very sensitive to changes in pressure and react to pressure exerted on the medium in which they are implanted/buried. A processor monitors/regulates the pressure inside the tubes and generates a signal if the pressure deviates from a determined norm.

When an intruder/vehicle approaches the detection zone, the ground starts to compress in direct relation to the extend of the pressure waves exerted by the weight and movement impact. The impact caused by a runner will create a greater pressure than a walker, a heavy person walking upright will create a greater pressure than a smaller person moving on hands and knees. The buried tube sensor nearest the point of pressure reacts to the energy (pressure) carried through the soil (buried medium) and in turn changes the pressure in the farthest tube proportionally. The pressure sensing unit detects the change in pressure in both tubes and generates an electrical signal proportional to the pressure exerted. The signals from both tubes are compared and transmitted to the analyzer. When the pressure between the two tubes exceeds a pre-set value, the analyzer generates an alarm signal.

NOTE: A self-compensating valve is used to maintain pressure within the tubes, adjusting to gradual/moderate changes associated with the burial medium such as those caused by moisture content (rain) or temperature changes (frost/drought). However, this valve does not adjust to rapid changes in pressure typical of personnel and vehicle movement, and other man-made or sudden natural movements such as earthquakes or explosions.

3.   Applications and Considerations:

    a.   Applications: The detection zone is created by burying the tubes approximately 4 feet apart, with the pressure-sensing unit linked and placed between the sensor tubes. Depending on the nature of the soil, this type of system can create a detection zone with up to a 350 feet radius. The depth at which the tubes are placed depends on the composition of the medium in which the tubes are placed. Normally, 10 inches is sufficient for earth and sand. Soil with an asphalt covering requires tubes to be placed at a more shallow depth of 4 - 8 inches. When working with a concrete surface/area, the sensor tubes should be buried just beneath the under surface of the concrete.

NOTE: Concrete is not a good conductor for the relative "light" pressure waves created by personnel, and in fact it serves as a good "insulator", thereby reducing the probability of human movement being detected. Therefore, it is essential to employ additional surveillance/detection means when dealing with expanses of concrete and possible human movement.

      b.   **Conditions for Unreliable Detection:** Because of the differential pressure principle employed and the nature of the self-compensating valve, the system has a high degree of immunity to typical environmental noise and weather conditions. However, tree roots closer than 10 feet to the sensor set can pose a problem due to the potential for windy conditions above ground can transfer pressure waves into the ground via the root system generating an alarm. Also, areas with heavy snowfall (and/or shifting sand) may have trouble with the system properly sensing seismic vibrations, depending on the depth and composition of the snow/sand.

      c.   **Causes for Nuisance Alarms:** Improper installation or calibration can cause normal activity to be interpreted as intrusion. Also, proximity to heavy road/rail traffic or seismic activity from pulsating or shock machinery can cause nuisance alarms.

4.   **Typical Defeat Measures:** Avoiding the potential zone(s) of detection; cushioning movement vibrations, dispersing/lowering impact energy, and/or bridging/planking over or through the detection zone are all, in varying degrees, viable defeat measures.

BALANCED BURIED PRESSURE LINE SENSOR

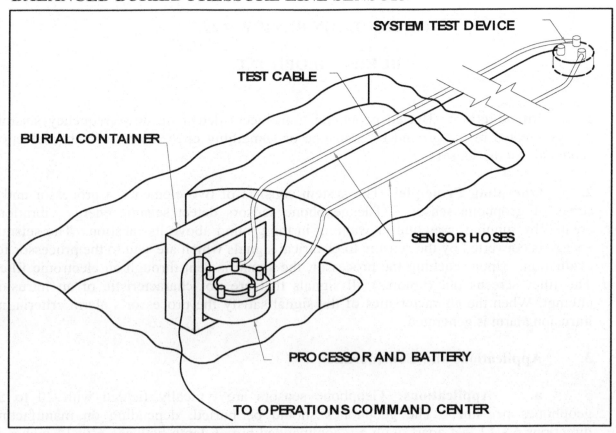

SYSTEM TEST DEVICE

TEST CABLE

BURIAL CONTAINER

SENSOR HOSES

PROCESSOR AND BATTERY

TO OPERATIONS COMMAND CENTER

TECHNOLOGY REVIEW # 25

BURIED GEOPHONE

1.   **Introduction:** Buried geophone transducers detect the low frequency seismic energy created in the ground by someone or something crossing through the detection screen above the sensors.

2.   **Operating Principle:** The system consists of two elements, a processor and a series of geophone sensors.   The geophone sensors detect seismic energy vibrations created by running, crawling or walking in the ground above its location.   The seismic energy is converted by the sensors to electrical signals which are sent to the processor for evaluation.   Upon reaching the processor, the signal is sent through an electronic filter. The filter screens out (ignores) all signals that are not characteristic of an intrusion attempt.   When the characteristics of the signal satisfy the processor's alarm criteria, an intrusion alarm is generated.

3.   **Applications and Considerations:**

    a.   **Applications:** Geophone sensors are typically fielded with 20 to 50 geophones per line.   The geophones should be buried, depending on manufacture directions 6 - 12 feet apart, with a recommended burial depth between 6 to 14 inches in soft to compact soil and 6 inches in asphalt.   It is recommended that burial field soil be stable and relatively compact, and the geophones should be installed between layers of sand, as compact sand is very conductive of seismic vibrations.   Geophone sensor zones lengths can extend up to 300 feet.

    An audio "listen-in" feature can be incorporated into the sensor field to aid in differentiating between nuisance alarms and valid intrusion attempts.   The listen-in feature allows the operator at a monitoring station to listen to the audible seismic signals from the geophones.   A trained operator can usually differentiate between normal stimuli and stimuli associated with intrusion attempts.

    b.   **Conditions for Unreliable Detection:** The main cause for unreliable detection is the burial medium in which the sensors are located.   Loose or inconsistent soil causes the seismic energy waves to have little effect on the geophones.

    c.   **Causes for Nuisance Alarms:** Geophones can detect very low levels of seismic activity, and because of this sensitivity, trees, fences, light poles, and telephone poles can pose major nuisance problems.   All of these items are anchored in the ground and transfer seismic energy to the ground when subjected to high wind.   Geophones should be installed at least 30 feet from trees, 10 feet from fences, and at a distance equal to the height of any nearby poles. Also, large animals passing over/through the detection zone can generate an alarm signal.

4.   **Typical Defeat Measures:** Bridging over the sensors will bypass the system.

BURIED GEOPHONE SENSORS

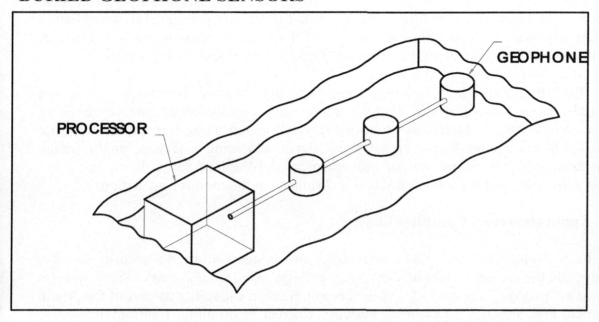

TECHNOLOGY REVIEW # 26

VIDEO MOTION DETECTION

1.   Introduction: Video (Image) Motion Detection (VMD) sensors use Closed Circuit Television (CCTV) systems (Visual, Low Level Light, and Infrared) to provide both an intrusion detection capability, and a means for security personnel to immediately and safely assess alarms (possible intrusions). CCTV systems provide the added benefit of documenting the events of an intrusion and the characteristics of the intruder.

2.   Operating Principle: Video Motion Detection sensors detect changes in the monitored area by comparing the "current" scene with a pre-recorded "stable" scene of the area. Video Motion Detectors monitor the video signal being transmitted from the camera. When a change in the signal is received, indicating a change in the image composition caused by some sort of movement in the field of surveillance, an alarm signal is generated, and the intrusion scene is displayed at the monitoring station.

3.   Applications and Considerations:

    a.   Applications: Once activated, most systems allow the security monitor to manipulate the camera's field of view, (e.g. enlarge, scan, tilt and pan). Some systems also have a "listening" , as well as a voice communication capability as part of the Alarm Assessment and Situation Monitoring system. Correct positioning, lighting conditions, and stability of cameras are all factors to be considered, as should striking a balance between the deterrent value of visible cameras and the security/monitoring value of concealed cameras. Both are valid applications.

    The installation configuration of a CCTV system is directly related to the nature of the security requirement.   Examples of monitoring capabilities include: dead zones between two fences, outside storage lots, interiors of warehouses (particularly at night), approaches to "rear doors", and vehicle/pedestrian entry points, loading docks and at guard posts where the CCTV system can be tied to a Duress Alarm. However, in all circumstances care must be given to securely mount the cameras, deny easy access to them, and keep the field of view as open and uncluttered as possible. In all applications, vegetation and obstacles to visual observation must be eliminated or reduced to a point where they do not detract from the utility of the system.

    b.   Conditions for Unreliable Detection: Areas that have poor lighting or extended periods of darkness may provide conditions for unreliable detection. Under these conditions both Infrared or Low Level Light camera configurations are recommended. Low light levels, even if the only source is ambient light, can be compensated for by the use of LLLTV cameras, whereas an infrared system is useful for detecting the "heat differential" generated by an intruder.

c.   **Causes for Nuisance Alarms:** When installing CCTV cameras, careful consideration must be given to the placement of the cameras to ensure the field of view will not be affected by: (1) natural light sources such as changes in the sun angle (sun rise/sun set) or scene brightness alterations from cloud motion, wind blown objects passing through the scene or camera vibrations, or (2) man-made light sources such as vehicle headlights, traffic lights, changes in parking lot or security lights patterns. Any of the above can generate an alarm signal, as each reflects as change in the image view. Insects flying close to the lens of the camera can also initiate an alarm signal and have been interpreted as larger objects moving in the field of coverage, however, a trained operator can detect this on the monitor.

4.   **Typical Defeat Measures:** An intruder aware of the system may be able to avoid detection by moving around the field of view. For this reason it is recommended that some of the cameras be placed as covertly as possible, and networked to one or more other sensors which can also act as a triggering or focusing mechanism.

TECHNOLOGY REVIEW # 27

RADAR

1. **Introduction:** Radar (<u>RA</u>dio <u>D</u>etection <u>A</u>nd <u>R</u>anging) is an active sensor that has undergone substantial refinement and enhancement since its first operational use as a detection sensor in the early 1940s. Radar uses ultrahigh frequency radio waves to detect intrusion of a monitored area.

2. **Operating Principle:** Radar sensors transmit a signal from an energy source in the ultrahigh frequency range of 100 MHz to 1 GHz. The Radar signal "bounces" off objects in the detection zone, and the reflected signal is then analyzed by a processor to determine the relative size, azimuth and distance of the object. The information is then converted to symbology and displayed as part of an integrated presentation on a local CRT (Cathode Ray Tube).

3. **Sensor Types/Configurations:** There are two basic types of radar sensors: monostatic sensors, which have the transmitter and receiver encased within a single housing unit, and bistatic sensors, in which the transmitter and receiver(s) are separate units creating a detection zone between/among them.

    a. **Monostatic Units:** In monostatic devices the transmitter and receiver are contained within one unit, referred to as a transceiver. Typically, detection for intrusion is achieved by the radar transceiver rotating in a pre-set "sweep" pattern. During rotation the transceiver transmits high frequency energy pulses, forming/scanning a detection zone. A signal processor, located within the transceiver, is programmed to recognize reflected energy from the normal environmental surroundings, thus not signaling an alarm. However, when a moving or foreign / new object is detected within the zone, a Doppler shift in the reflected energy is created. When the magnitude of the reflected energy surpasses the processors criteria, an alarm signal is generated.

    b. **Bistatic Units:** The transmitter and receiver(s) for bistatic models are separate units. The detection zone is created between the units. The transmitter is typically transmitting in a designated "sweep" pattern, with receivers at several locations designed to maximize the potential for detection. The transmitter generates a field of high frequency energy, which "bounces/reflects" off "foreign" objects and is detected by one or more receivers. When the resulting signals satisfy the detection criteria, an alarm signal is generated.

4. **Applications and Considerations:**

    a. **Applications:** Radar sensors are used primarily to monitor exterior areas, although in some situations Radar sensors can be used to monitor large interior open areas. In both situations, the ground should be reasonably level and the perimeter boundaries straight. If portions of the perimeter are hilly or have crooked boundaries, the radar unit may be elevated to provide a better line of sight/view, or radar sensors can be used to monitor the straight and level sections of the perimeter, while other types of detectors (e.g. In-ground sensors, video motion detection) can be used to monitor the remaining sections. The use of a companion system, such as video image motion detection not only provides a second line of defense, but it provides security personnel with an additional tool to assess alarms and discriminate actual/potential penetrations from false alarms or nuisance events.

NOTE: Radar sensors can also be very useful in detecting plane or helicopter-borne intrusion attempts, which would otherwise bypass ground-oriented, perimeter sensors (e.g., fences and in-ground).

       b.   **Conditions for Unreliable Detection:** "Dead zones" created by large objects, buildings or hill masses/depressions can provide safe havens for intruders, allowing them to avoid the radar field. In addition, extreme weather conditions, such as rain/snow storms can decrease detection potential.

       c.   **Causes for Nuisance Alarms:** Nuisance alarms can be generated by detection of foreign objects outside the protected area or by the random reflection of radar energy.

5.   **Typical Defeat Measures:** Uneven terrain may create enough "hidden pockets", allowing the intruder to be undetected by using a slow/low approach pattern through the protected volume.

TECHNOLOGY REVIEW # 28

ACOUSTIC DETECTION (AIR TURBULENCE)

1.   Introduction: Acoustic air turbulence sensors detect low frequencies created by helicopters that are in their final landing phase or at close range (1 to 2 miles). This sensor can be very useful in detecting helicopter-borne intrusion attempts, which would otherwise bypass normal perimeter sensors (fence and in-ground).

2.   Operating Principle: Acoustic air turbulence sensors "listen" for basic sound pressure waves generated by helicopter rotor blades. The helicopter has four main acoustic sources for producing these waves: 1) downward wash, caused by the energy that is required by the rotor blades to keep airborne; 2) blade slap, which originates from the forward traveling blade as it penetrates the trailing tip vertex remaining from the previous blade passage; 3) tail rotor, which is directly geared to the main rotor that generates harmonically-related frequencies, and 4) engine noise, which is not usually muffled because of the power needed for rapid changes in flight performance. With these acoustic sources the approaching helicopter will produce frequencies within a range of 20 - 40 Hz, depending on the model. Once frequencies are detected, the acoustic air turbulence sensor sends the signal to a processor that filters out frequencies not associated with helicopter flight. If the signal passes through the narrow acoustic band filter of 20 - 40 Hz, an alarm signal is generated.

3.   Applications and Considerations:

    a.   Applications: Under test conditions some helicopters, including some of the quietest, such as the Hughes 500C, Bell 47, Bell 206, and the Jet Ranger, have been detected at distances up to 500 feet. However, for increased probability and reliability of detection, detector sensitivity is typically set for a range of 300 feet. Detection zones should overlap to insure all approach/segments of the protected area are covered by at least one sensor. The sensors should be located away from any vehicular/road traffic and/or railroad right of ways to minimize potential interference from any pressure/wind turbulence generated by high speed truck or train movement. There is no restriction on the distance of sensors from the main control unit, as long as system communication is properly designed. Therefore, the protected area can literally extend for miles.

    b.   Conditions for Unreliable Detection: Sensitivity settings set too low to compensate for vehicular traffic. Also, improper spacing of the sensors, allowing some areas not to be covered by the detection pattern.

    c.   Major Causes for Nuisance Alarms: Wind generating a broad band noise causes the most difficulty for the acoustic air turbulence sensor. Also turbulence generated at a distance and conveyed via pressure wave propagation can be interpreted as broad band rumbling if they impact the pressure sensitive transducer of the sensor.

Turbulence created by wind viscosity and the roughness of the terrain can also generate conditions for nuisance alarms

4. **Typical Defeat Measures:** The system will not detect air borne assault of other methods, such as glider, parachute, or ultralight.

ACOUSTIC / AIR TURBULENCE SENSOR

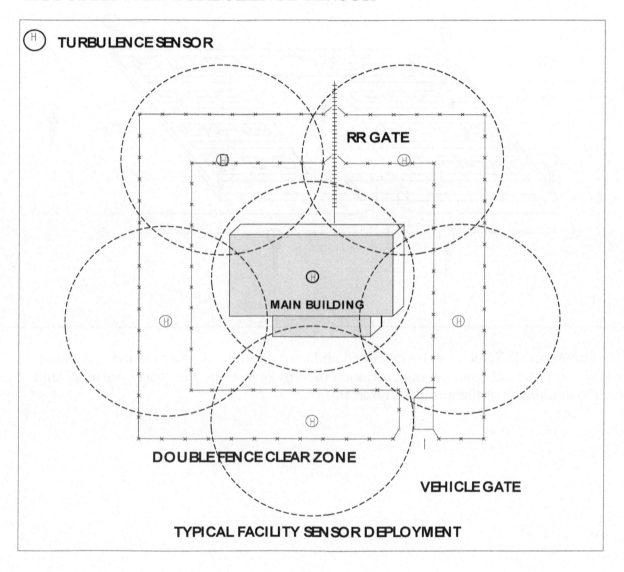

LOGISTICS / MUNITIONS STORAGE COMPLEX

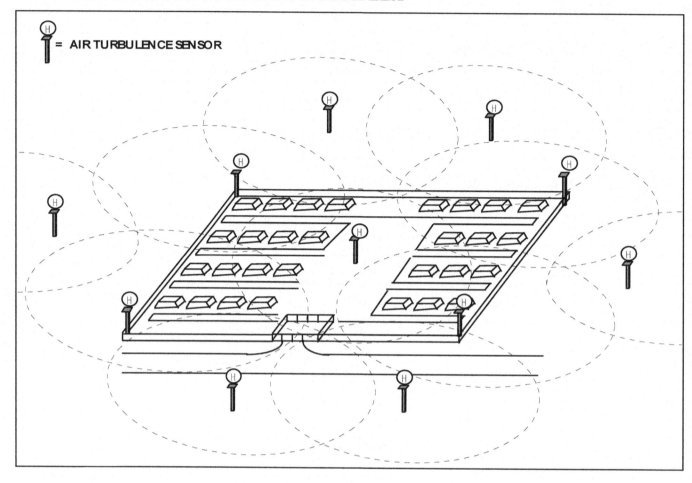

= AIR TURBULENCE SENSOR

NOTE: The Acoustic Screen can be extended and configured to conform to the surrounding topography and potential approaching corridors, thereby increasing the "early warning" and positional location/axis of the intruding aircraft.

CPSIA information can be obtained
at www.ICGtesting.com
Printed in the USA
BVOW04s2113031217
501844BV00006B/29/P